Mina Nishi

Laminar to turbulent transition in pipe flow through puffs and slugs

Mina Nishi

Laminar to turbulent transition in pipe flow through puffs and slugs

Detailed experimental investigations of transitions in low and high Reynolds number pipe flows

Südwestdeutscher Verlag für Hochschulschriften

Impressum/Imprint (nur für Deutschland/ only for Germany)
Bibliografische Information der Deutschen Nationalbibliothek: Die Deutsche Nationalbibliothek verzeichnet diese Publikation in der Deutschen Nationalbibliografie; detaillierte bibliografische Daten sind im Internet über http://dnb.d-nb.de abrufbar.

Alle in diesem Buch genannten Marken und Produktnamen unterliegen warenzeichen-, marken- oder patentrechtlichem Schutz bzw. sind Warenzeichen oder eingetragene Warenzeichen der jeweiligen Inhaber. Die Wiedergabe von Marken, Produktnamen, Gebrauchsnamen, Handelsnamen, Warenbezeichnungen u.s.w. in diesem Werk berechtigt auch ohne besondere Kennzeichnung nicht zu der Annahme, dass solche Namen im Sinne der Warenzeichen- und Markenschutzgesetzgebung als frei zu betrachten wären und daher von jedermann benutzt werden dürften.

Verlag: Südwestdeutscher Verlag für Hochschulschriften Aktiengesellschaft & Co. KG
Dudweiler Landstr. 99, 66123 Saarbrücken, Deutschland
Telefon +49 681 37 20 271-1, Telefax +49 681 37 20 271-0, Email: info@svh-verlag.de
Zugl.: Erlangen, Uni, Diss., 2009

Herstellung in Deutschland:
Schaltungsdienst Lange o.H.G., Zehrensdorfer Str. 11, D-12277 Berlin
Books on Demand GmbH, Gutenbergring 53, D-22848 Norderstedt
Reha GmbH, Dudweiler Landstr. 99, D- 66123 Saarbrücken
ISBN: 978-3-8381-0318-1

Imprint (only for USA, GB)
Bibliographic information published by the Deutsche Nationalbibliothek: The Deutsche Nationalbibliothek lists this publication in the Deutsche Nationalbibliografie; detailed bibliographic data are available in the Internet at http://dnb.d-nb.de.

Any brand names and product names mentioned in this book are subject to trademark, brand or patent protection and are trademarks or registered trademarks of their respective holders. The use of brand names, product names, common names, trade names, product descriptions etc. even without
a particular marking in this works is in no way to be construed to mean that such names may be regarded as unrestricted in respect of trademark and brand protection legislation and could thus be used by anyone.

Publisher:
Südwestdeutscher Verlag für Hochschulschriften Aktiengesellschaft & Co. KG
Dudweiler Landstr. 99, 66123 Saarbrücken, Germany
Phone +49 681 37 20 271-1, Fax +49 681 37 20 271-0, Email: info@svh-verlag.de

Copyright © 2008 Südwestdeutscher Verlag für Hochschulschriften Aktiengesellschaft & Co. KG and licensors
All rights reserved. Saarbrücken 2008

Produced in USA and UK by:
Lightning Source Inc., 1246 Heil Quaker Blvd., La Vergne, TN 37086, USA
Lightning Source UK Ltd., Chapter House, Pitfield, Kiln Farm, Milton Keynes, MK11 3LW, GB
BookSurge, 7290 B. Investment Drive, North Charleston, SC 29418, USA
ISBN: 978-3-8381-0318-1

Laminar to turbulent transitions in pipe flow through puffs and slugs

Der Technischen Fakultät der
Friedrich-Alexander-Universität
Erlangen-Nürnberg

zur Erlangung des Grades

DOKTOR-INGENIEURIN

vorgelegt von
Mina Nishi

Erlangen, 2009

Als Dissertation genehmigt von der Technischen Fakultät
der Universität Erlangen-Nürnberg

Tag der Einreichung: 02.02.2009
Tag der Promotion: 15.06.2009

Dekan: Prof. Dr. Dr. habil. Johannes Huber
Berichterstetter: Prof. Dr. Dr. h.c. Franz Durst
                    Prof. Dr. Bruno Eckhardt

Abstract

A classical problem in fluid mechanics, laminar to turbulent transitions in pipe flow, was investigated and the results of new findings are summarized in this thesis. The investigation of transitions in pipe flows was started by Reynolds (1883) and continued subsequently, hence it has a long history. Nevertheless, there are still many unknown issues that need clarification, which are presented as an introduction containing a brief literature survey. For the actual experimental investigations, an experimental test rig, built by Durst & Ünsal (2006), was applied after some modifications. The test rig, containing a mass flow controller, a flow conditioner, a brass pipe, devices to trigger transition and also measuring equipment, is described in detail. Through some preliminary research work, a natural transition and also a transition triggered by ring obstacles were studied. The investigation showed clearly the dependence on the height of the ring obstacles and also on Reynolds number that a flow was triggered to a transitional flow. The Reynolds number was defined as $Re = u_{bulk} D/\nu$, where u_{bulk} is the bulk velocity, D is the pipe diameter and ν is the kinematic viscosity of the fluid. As is well known, two distinct types of flow structure exist during a laminar to turbulent transition, namely puffs and slugs, first introduced by Wygnanski & Champagne (1973). The test rig was then applied to investigate the development of puffs and slugs which were triggered in a repeatable way in a pipe at different Reynolds numbers. Also considered is a transformation from puffs to slugs through puff splittings, occurring while they propagate in a pipe or occurring by an increase in Reynolds number.

There are other interesting issues observed in low Reynolds number pipe flows, such as the dissipation of puffs. The time between when puffs were created and dissipated is named 'full-lifetime' in this thesis. The full-lifetime of puffs was measured directly with a pressure transducer and the results are presented and discussed. Through the investigation of lifetime, the possible evolutions of flow structures occurring in laminar to turbulent transitions observed in low Reynolds number pipe flows could be well explained. Finally, for further understanding of transitional phenomena in pipe flows, the

Reynolds stress anisotropy invariant of the slug structure was carefully measured by using hot wire anemometry. The results indicated that the Reynolds stress anisotropy model is able to predict the transition.

Conclusions and final remarks are presented, summarizing all the investigations carried out.

Contents

Abstract   3

Contents   5

Nomenclature   6

1 Introduction and brief literature survey   9
  1.1 Beginning of the study of transitions in pipe flows 10
  1.2 Critical Reynolds number 14
  1.3 Puffs and slugs in transitional pipe flows 17
  1.4 Decay of puffs and traveling waves 19
  1.5 Contents of the thesis 23

2 Test rig for the investigations   27
  2.1 Mass flow rate control system 27
  2.2 Flow conditioner and pipe test rig 33
  2.3 Devices for triggering laminar to turbulent transitions 34
  2.4 Measurement equipment and data acquisition system 37

3 Preliminary research work   40
  3.1 Development of laminar flow and natural transitions 41
  3.2 Transitions triggered by ring obstacles 46
  3.3 Triggering of fully developed laminar flow 53
  3.4 Large pipe test rig transitions triggered by ring obstacles . . . 56
  3.5 Summary of natural and ring obstacle-triggered transitions . . 59

4 Development of puffs and slugs   62
  4.1 Slug development in a pipe 63
  4.2 Puff development in a pipe 69
  4.3 Puff to slug transformation 75
  4.4 Summary of development of puffs and slugs 82

5 Lifetime of transitional flow structures   85
  5.1 Method of direct measurement of lifetime 86
  5.2 Analysis of lifetime and probability measurements 91

	5.3 Summary of the results	100
6	**Conclusions**	**103**
7	**Outlook for future investigations**	**106**
	7.1 Anisotropy-invariant mapping	107
	7.2 Reynolds stress anisotropy measurement of slugs	109
	7.3 Conclusions of the anisotropy measurement of slugs	110
	References	**113**
	Kurzfassung	**117**
	Acknowledgements	**120**

Nomenclature

a	coefficient(1) for a hot wire calibration
A	amplitude of disturbances
b	coefficient(2) for a hot wire calibration
c_f	coefficient of pipe loss
c_p	gas specific heat at constant pressure
c_v	gas specific heat at constant volume
d, D	pipe diameter
E	signal potential of a hot wire anemometer
F	area of a valve opening
h	obstacle height of ring and iris diaphragm
h^+	non-dimensionalized height of ring obstacle
L	pipe length
LT	living time
n	ordinal number of sampling
$\dot{M}$	mass flow rate
N	quantity and total number of sampling
N_0	initial quantity at time $t=0$
P_H	pressure of higher pressure chamber
P_L	pressure of lower pressure chamber
$P_{\tau(*)}$	probability of occurrence of puffs
R	pipe radius
$\mathbf{R}$	gas constant
Re	Reynolds number based on the pipe diameter $(uD\nu^{-1})$
Re_{crit}	critical Reynolds number

Re_{min}	minimum critical Reynolds number
Re_λ	Reynolds number based on the length scale of the disturbance
Re_τ	Reynolds number based on u_τ $(u_\tau D \nu^{-1})$
t	time
t_{turb}	turbulent duration time
T	total measurement time
T_H	temperature of higher pressure chamber
Tu	turbulent intensity $(u'/\bar{u})$
u	instantaneous velocity
$\bar{u}$	time-averaged velocity
$\tilde{u}$	realization-averaged velocity
u'	rms value for u
u_{bulk}	bulk velocity
u_{prop}	propagation velocity
u_τ	friction velocity $(\tau_w/\rho)^{1/2}$
U	velocity of flows

Greek letters

α	decay constant
Δp	pressure difference
Δt_s	duration of slug
ϵ	factor of critical Reynolds number to a disturbance amplitude
ϕ	hole diameter
γ	intermittency factor
κ	heat capacity ratio c_p/c_v
λ	Taylor micro scale
ν	kinematic viscosity of fluid
θ	mass flow density
ρ	density of fluid
τ	lifetime
$\tau_{1/2}$	half-life
τ_w	wall shear stress
τ^*	non-dimensional length L/D

Chapter 1

Introduction and brief literature survey

Studies of transitions in pipe flows have a long history after the well-known work of Reynolds (1883). Recently, Willis *et al.* (2008) mentioned in their review paper on the progress in investigations of pipe flow transitions, 'The process whereby turbulence arises is still not understood even in outline', as the problem has become one of the outstanding challenges of fluid mechanics. As is reported, there are many unknown issues related to transitions in pipe flows such as their origins, critical Reynolds numbers (the minimum Reynolds number at which a transitional flow can be sustained, or the maximum Reynolds number at which a flow can remain as a laminar flow), a minimum amplitude of disturbances to trigger transitions at different Reynolds numbers, the development of flow structures of laminar to turbulent transitional flows including necessary evolution distances for triggered flow structures into sustained flow structures and decay of transitional flow structures especially at low Reynolds number. Additionally, the dependences of different types of disturbances on all the above issues are not yet well described. One of the reasons why transitional phenomena in pipes are not fully understood, although their study has had over 125 years of history, could be that it is widely accepted that a fully developed laminar flow is linearly stable to infinitesimal perturbations (e.g. Meseguer & Trefethen (2003)); however, at a high Reynolds number, as in most industrial pipe flows, flows in pipes appear as turbulent flows and one observes transitions in pipes even at very low Reynolds number. It is also observed that transitions in pipes depend not only on Reynolds number but also on the amplitude of an existing disturbance and on the sensitivity of the flows.

In this chapter, literature related to the thesis topic is briefly surveyed as an introduction to the study of laminar to turbulent transition in pipe flows. First, the beginning of studies on them is described. Then the investigations are categorized and described in three separate sections devoted to studies on critical Reynolds number (section 1.1), on transitional flow structures, i.e. puffs and slugs, the appearance of which depends on Reynolds number (section 1.2), and on the decay of transitional flow structures (section 1.3). The overall contents of the thesis are described in the last section (section 1.4).

1.1 Beginning of the study of transitions in pipe flows

Reynolds (1883) considered that flow characteristics, i.e. laminar or turbulent flow, depend on different physical properties and fluid parameters, such as the bulk velocity of the pipe u_{bulk}, the pipe diameter D and the kinematic viscosity of the fluid ν. He showed through his experiments, using the test rig shown in figure 1.1, that a non-dimensional number, which today is called the Reynolds number ($Re = u_{bulk}D/\nu$), determines flow characteristics. The setup in figure 1.1 was used for visualization of laminar and turbulent pipe flow, the flow rate being controlled by the valve fitted next to the pipe test section. The visualization of pipe flows was carried out using colored water, and the results shown in figure 1.2, characterize the difference between laminar and turbulent (transitional) flow. Reynolds (1883) also mentioned that a development length in the pipe was required for the flow to change from a laminar to a turbulent state, depending on the Reynolds number, and also the laminar to turbulent transition in pipe flows first occurred intermittently, as sketched in his paper, and also shown in figure 1.2.

Taking a constant pipe length, one observes that the turbulent intermit-

Figure 1.1: A sketch of the experimental setup of Reynolds (1883)

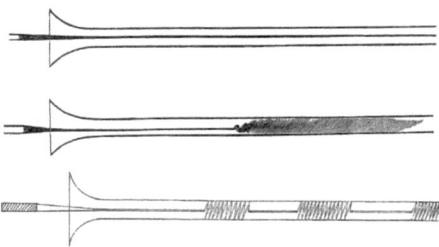

Figure 1.2: Sketches of laminar and turbulent (transitional) flow (top and middle) and an intermittent nature of a transitional flow (bottom) in pipes, as shown by Reynolds (1883)

tency of a flow varies from 0 to 1 over a small but finite Reynolds number range, where the turbulent intermittency was defined first by Rotta (1956) as the ratio of the sum of the time of turbulent flows divided by the total measurement time. Over this length, the flow loses its strongly intermittent nature and turns into a fully developed turbulent pipe flow.

Rotta (1956) developed a sophisticated experimental pipe test rig for investigating transitional flows from laminar to turbulent by employing a critical nozzle. The principle of the critical nozzle is governed by a gas dynamic equation as follows:

$$\dot{M} = \left(\frac{2}{1+\kappa}\right)^{\frac{1}{\kappa-1}} F_D \sqrt{\frac{2\kappa}{1+\kappa} P_D \rho_D} \qquad (1.1)$$

where $\dot{M}$ is the mass flow rate, κ is the ratio of the specific heat coefficients, F_D is the opening area and P_D and ρ_D are the pressure and density of the fluid at the nozzle inlet, respectively. Equation 1.1 shows that $\dot{M}$ depends on κ, F_D, P_D and ρ_D. If the pressure at the outlet of the nozzle is lower than the critical pressure P_{crit}, defined in the following equation, $\dot{M}$ is choked, hence it is independent of the pressure change of the lower pressure side (outlet of the nozzle):

$$P_{crit} = \left(\frac{2}{1+\kappa}\right)^{\frac{1}{\kappa-1}} \approx 0.53 P_D. \qquad (1.2)$$

It was considered that for investigating transitional flows, the mass flow must be kept constant by such a method, the principle of which was mentioned above, since the pressure drops suddenly in the pipe when a transition in pipe flow occurred. Rotta (1956) employed hot wire anemometry to measure the velocity-time profile at the exit of the pipe, and the results obtained are shown in figure 1.3. As the Reynolds number increases, turbulent flows appear more frequently, thus the turbulent intermittency increases with increase in Reynolds number, following on S-shaped curve. It was also reported that the turbulent intermittency factor increased with increase in pipe length (x/d in figure 1.3).

Numerous experimental, numerical and theoretical studies of laminar to turbulent transitions in pipe flows have been carried out since the work of Reynolds (1883) and Rotta (1956), and these are introduced in the following sections.

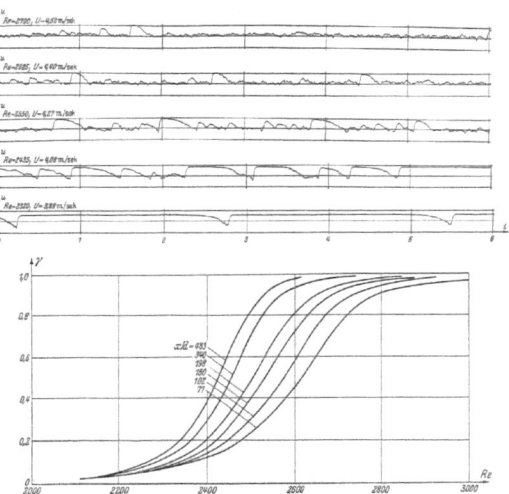

Figure 1.3: Velocity-time profile measured by hot wire anemometry at the exit of the pipe (top) and turbulent intermittency factor (γ) curve shown for different pipe lengths (x/d, where x is the distance from the inlet and d is the pipe diameter) (bottom), as shown by Rotta (1956)

1.2 Critical Reynolds number

There have been a large number of discussions regarding critical Reynolds number, which is still one of the most intriguing issues in the investigations of transitions in pipe flows. Reynolds (1883) reported that there were two kinds of critical Reynolds numbers: one a Reynolds number above which a turbulent (transitional) flow can be observed and the other a Reynolds number below which no transitional flow structures could be sustained. The former critical Reynolds number is, theoretically, infinitely high as e.g. Meseguer & Trefethen (2003) showed that Hagen-Poiseuille flow was stable to infinitesimal perturbations. Experimentally, Pfenninger (1961) was able to maintain the pipe flow in a laminar state up to $Re = 10^5$. Although a fully developed laminar flow should be stable toward any disturbances, it turns into a turbulent flow above its critical Reynolds number due to certain conditions.

Wygnanski & Champagne (1973) attempted to find the relationship between disturbance amplitude and critical Reynolds number, with the results shown in figure 1.4. This figure shows that a finite amplitude of disturbances is required to trigger the flow from laminar to turbulent. The required amplitude became smaller with increase in Reynolds number. It is clear from figure 1.4 that there is a nonlinear relation at low Reynolds number between the disturbance amplitude and the critical Reynolds number, although with increase in the disturbance amplitude, the critical Reynolds number remains more or less the same as the amplitude of the dependence of disturbance on Reynolds number is different for below and above $Re \approx 2700$. Figure 1.4 suggests that below a threshold Reynolds number, transitional flow structures decay in a pipe, although any large-amplitude disturbance is applied. It is said that such decays occur within a distance $100D$ from the triggering location, where D is the pipe diameter. The decay of transitional flow structures (i.e. a turbulent to laminar transition) is described with the introduction of previous studies in section 1.4.

Darbyshire & Mullin (1995) applied various disturbances such as jet injection and suction in a fully developed pipe flow to create repeatable disturbances. They reported that the minimum amplitude of disturbance for triggering the flow decreased with increase in Reynolds number but remained

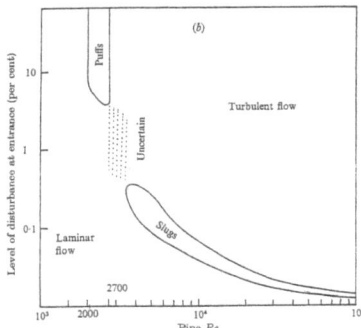

Figure 1.4: Nonlinear relation of disturbance amplitude and critical Reynolds number shown by Wygnanski & Champagne (1973)

finite at $Re = 10^4$, as shown in figure 1.5, and were relative insensitive to the form of disturbances. Draad et al. (1998) applied a porous pipe section to apply periodic suction and injection for triggering flows and found the correlation between the minimum amplitude of the disturbance and the critical Reynolds number to be $A \propto Re_{crit}^{-1}$ for high-frequency disturbances and $A \propto Re_{crit}^{-1.5}$ for low-frequency disturbances, where A is the amplitude of the disturbance and Re_{crit} is the critical Reynolds number. Experimental investigations were further carried out to find the exponential factor for the minimum amplitude of disturbances, which is represented by ϵ in the equation $A \propto Re_{crit}^{\epsilon}$ according to e.g. Hof et al. (2003) and Peixinho & Mullin (2007). Hof et al. (2003) reported that they found $\epsilon = -1$ as a constant value even when different jet disturbances (various injection quantities and durations) were applied. Peixinho & Mullin (2007) applied jet and push-pull disturbances, which was intended to avoid a global mass flow change, and reported $\epsilon = -1$ for jet and $\epsilon = -1.3$ to -1.5 depending on the orientation with respect to the stream-wise direction of the disturbances. The discrepancies in ϵ could be caused by different definitions of disturbance amplitudes, as Trefethen et al. (2000) pointed out.

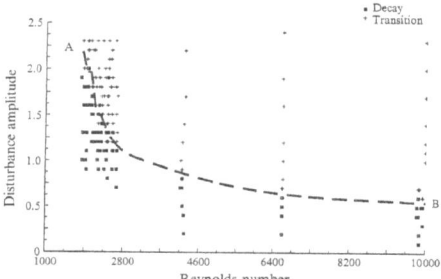

Figure 1.5: Correlation between the amplitude of a disturbance and the critical Reynolds number obtained by Darbyshire & Mullin (1995)

One of the difficulties in numerical investigations of transitions of a pipe flow is to relate the physical and numerical disturbance amplitudes to each other. However, Shan et al. (1998), Meseguer & Trefethen (2003), Eckhardt & Mersmann (1998) and Mellibvosky & Meseguer (2006) carried out direct numerical simulations of pipe flows with periodic boundary conditions and reported $\epsilon \approx -1$ to -1.5. To find the exponential factor of the Reynolds number is of special interest as it relates to the issue of whether there is a finite amplitude of disturbance to trigger a flow into a turbulent state as the Reynolds number increases towards infinity, however, the ultimate factor has not yet been found (cf. Trefethen et al. (2000)).

There is another threshold Reynolds number, pointed out by Reynolds (1883), below which no triggered flow structures could be sustained. Jovanović & Pashtrapanska (2004) carried out a theoretical analysis by applying a Reynolds stress anisotropy model and found that at $Re \leq 1930$ no transitional structures can be sustained in pipe flows if the applied disturbance has two components. Willis & Kerswell (2007) carried out numerical and theoretical analyzes and concluded that at $Re = 1870$ the transitional flow structures suddenly become attractors. Nevertheless, Darbyshire & Mullin (1995) showed the existence of a transitional structure for $Re \geq 1760$. Even at $Re < 1760$, it is still possible to observe transitional flow structures at a location close to the pipe inlet. Since it has been considered that triggered flows need at least $100D$ to develop into isolated transitional flow patches (puffs) in low Reynolds number pipe flows, the evolution of triggered structures in such short lengths from the triggering location has not been well investigated. Additionally, the boundary between 'just triggered' flows and puffs is not yet well described and therefore, how much time and space 'just triggered' flows take to evolve into puffs is not yet clear. The issue related to the transitional flow structures at low Reynolds number, including their decay, is introduced with a literature survey in section 1.4.

1.3 Puffs and slugs in transitional pipe flows

Wygnanski & Champagne (1973) first introduced that there are two different flow structures, observed in a laminar to turbulent transition in pipe flows, which are known as puffs (flow structures observed at low Reynolds number) and slugs (flow structures observed at high Reynolds number). Both puffs ans slugs are isolated patches of flow structures, hence the boundary between a laminar and a puff or a slug flow can be recognized. Whether one or the other structure occurs is dependent not only on the Reynolds number but also on the flow disturbance amplitude introduced, as shown in figure 1.4. Wygnanski & Champagne (1973) developed a pipe test section combined with an air blower through a contraction nozzle. They employed an orifice plate as a disturbance to trigger the flow and carried out hot wire anemometry measurements to show gross characteristics of slugs and puffs. Wygnanski & Champagne (1973) indicated that the transitional structures appear to be slugs at $Re \gtrsim 2500$ and slug structures have clear boundaries on both the front and back edges between laminar and turbulent flows. On the other hand, puffs have a clear boundary only at their back edge and the front edge is not very clear. Lindgren (1969) and Wygnanski & Champagne (1973) showed that the front and back edges have different propagation velocities as the propagation of the front edge was faster and the back edge was slower than the bulk velocity, so that the extent of the slugs increases as they move downstream in a pipe. Furthermore, Wygnanski & Champagne (1973) mentioned that slug structures have within themselves turbulence properties similar to those of fully developed turbulent flow if the Reynolds number is sufficiently high.

Subsequently, Wygnanski *et al.* (1975) studied puffs at $2000 \leq Re \leq 2700$. They initiated the puffs using a 1 mm diameter jet produced by a driven loudspeaker and emanating from the wall at the pipe entrance and described the internal structure of puffs clearly. They observed puffs at a certain Reynolds number propagated downstream with a velocity close to the bulk velocity without changing their form, hence they named them equilibrium puffs. They reported that a puff extends from 5 to 20 pipe diameters, depending on Reynolds number, triggering conditions and how far it propagates downstream. They showed the interior of puffs in detail based on the velocity measurements.

Wygnanski *et al.* (1975) also observed an interesting phenomenon in transitions in pipe flows, namely puff splitting, which occurred in certain range of Reynolds number. They showed how the puff splitting occurrence changes with increase in Reynolds number and that when puff splitting occurred, the length of the puff suddenly increased.

The shortcoming of jet triggering of puffs, i.e. a large global or local variation in the mass flow rate during the operation of the large-amplitude disturbance, was removed by using orifice plates or ring obstacles at the inlet of the pipe by Rubin *et al.* (1980), who showed that puffs were independent of the type of triggering device and its stream-wise location, and Bandiopadhyay (1986), who visualized puffs to show their structural characteristics. Durst & Ünsal (2006) investigated structures in laminar to turbulent transitional flows by performing hot wire velocity measurements at the end of an $L/D = 666.7$ long pipe, where L is the pipe length and D is the pipe diameter. Through ring obstacles of different heights h, placed at the pipe inlet, the critical Reynolds number was decreased from 11500 to approximately 2300 in their case, where the transitional flow structures were formed as puffs for $2000 \lesssim Re \lesssim 2500$ and in the form of slugs for $3000 \lesssim Re \lesssim 11500$. They measured the propagation velocity of puffs and slugs, created by a sophisticated triggering system with an iris diaphragm, through pressure measurements over the entire pipe length.

Although the Reynolds number range in which puffs and slugs would appear is already clear, the turbulent intermittency varies at different Reynolds numbers depending on the pipe length. Additionally, puff splitting is an interesting bifurcation phenomenon in the nature of transitional pipe flows which was not investigated further after having been reported first by Wygnanski *et al.* (1975). These issues are closely related to the development of puffs and slugs in a pipe and ought to be investigated to describe the transitional phenomena in pipe flows in more detail.

1.4 Decay of puffs and traveling waves

Willis *et al.* (2008) mentioned in their review paper, summarizing experimental, numerical and theoretical investigations of pipe flow transitions, that progress has been made with three fundamental issues: the threshold amplitude of disturbances required to trigger a flow from a laminar to a turbulence state, the threshold Reynolds number below which a disturbance decays from a turbulent to a laminar state and the relevance of recently discovered families of unstable traveling wave solutions to transitional and turbulent pipe flow. The first issue has already been introduced in section 1.2 and the second and third issues are introduced here with a literature survey.

Darbyshire & Mullin (1995) observed that transitional flow structures can decay (relaminarize) at very low Reynolds number in constant mass flow rate pipe flows. The phenomenon of decaying transitional structures in low Reynolds number pipe flows was then studied by e.g. Faisst & Eckhardt (2004), Mullin & Peixinho (2006), Peixinho & Mullin (2006), Hof *et al.* (2006) and Willis & Kerswell (2007). Once a transitional flow structure has been created by a triggering, it convects and may decay in the downstream region of the pipe. Accordingly, a measurement of the probability (rate of existence) of transitional flow structures at the exit of an arbitrary length pipe can determine that the transitional flow structure survived at least for the pipe length and a time t that a transitional flow structure needs to propagate from the inlet to the pipe outlet.

Mullin & Peixinho (2006) measured the probability of transitional flow structures at several distances from the triggering location, and their results are shown in figure 1.6. They obtained a linear fitting curve by the least squares method, showing the relation between the probability decrease with increase in the distance from the triggering location, as shown in figure 1.6(a) and (c). Sreenivasan (1982) predicted that the disturbed flow decays exponentially and Mullin & Peixinho (2006) showed experimentally that the probability of the transitional structures decreases exponentially with increase in distance. A quantity is said to be subject to exponential decay if it decreases at a rate proportional to its value. This can be expressed symbolically by the equation $dN/dt = -\alpha N$, where N is the quantity and α is a positive number called the decay constant. The solution to this equation is, $N(t) = N_0 e^{-\alpha t}$, where $N(t)$ is the quantity at time t and $N_0 = N(0)$ is the initial quantity at time $t = 0$. It is considered that it is possible to compute the average length of time for which an element remains in the set if the decaying quantity is the number of discrete elements of a set. Such a time is called the lifetime and it can be shown that it relates to the decay rate, $\tau = 1/\alpha$. The lifetime is thus seen to be a characteristic time as $N(t) = N_0 e^{-t/\tau}$. Hence, it is the time needed for the assembly to be reduced by a factor e. When the base of the exponential is chosen as 2 instead of e, the scaling time is called the half-life. A half-life, which for the convenience is written here as $\tau_{1/2}$, is the time required for the exponentially decaying quantity to fall to half of its initial value and can be written in terms of the decay constant α or the lifetime τ as $\tau_{1/2} = \ln 2/\alpha = \tau \ln 2$. When this expression is inserted for τ in the exponential equation above, and ln2 is absorbed into the base, this equation becomes $N(t) = N_0 2^{-t/\tau_{1/2}}$. Thus the amount of material left is $2^{-1} = 1/2$ raised to the number of half-lives that have passed. Accordingly, Mullin & Peixinho (2006) calculated the half-life from the probability measurement, which they denoted τ as shown in figures 1.6(b) and (d).

Hof et al. (2006) also obtained experimentally the lifetime of transitional structures based on the probability change with time scale, and numerically Faisst & Eckhardt (2004), Hof et al. (2006) and Willis & Kerswell (2007) obtained lifetimes with the results shown in figure 1.7. Faisst & Eckhardt (2004) obtained results which suggested that the lifetime of transitional structures would converge to a finite value. Hof et al. (2006), who carried out experimental and numerical investigations, also concluded that the lifetime converges with increase in Reynolds number. On the other hand, Mullin & Peixinho

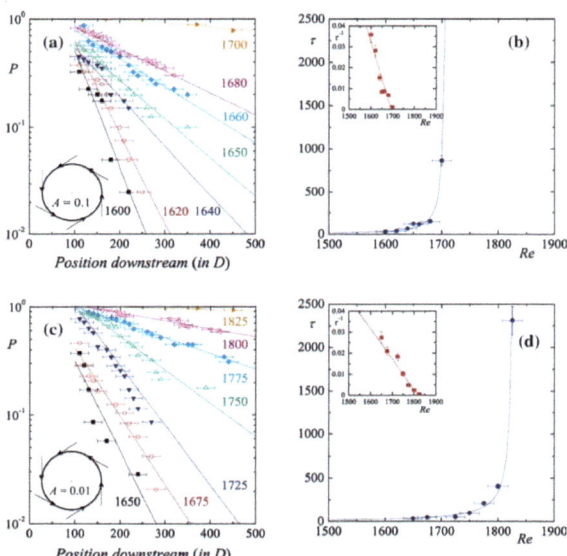

Figure 1.6: Probability of transitional structures at different Reynolds number (a and c) and half-life τ (b and d), measured by Mullin & Peixinho (2006)

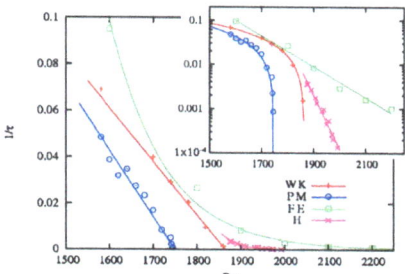

Figure 1.7: Lifetime results obtained by Faisst & Eckhardt (2004) (FE), Peixinho & Mullin (2006) (PM), Hof *et al.* (2006) (H) and Willis & Kerswell (2007) (WK)

(2006) applied different amplitudes of disturbance to trigger the flow and reported that the lifetime dependences differed (cf. figure 1.6(b) and (d)); however, in principle, their results indicated that the lifetime would diverge at a finite Reynolds number. Willis & Kerswell (2007) applied direct numerical simulation and found that $Re = 1870$ is a 'crisis Reynolds number' at which the lifetime of the transitional structure started to diverge.

Additionally, Peixinho & Mullin (2006) first kept the Reynolds number and initial conditions constant so that all transitional structures created by triggering device were similar to each other, and then decreased the Reynolds number suddenly down to predetermined values (Re_{low}) in order to observe whether the created transitional structures decayed or survived in a certain downstream region of the pipe. They found that at $Re_{low} = 1750$, as the final decreased Reynolds number, the transitional structures showed a critical behavior such that they started to have diverging lifetimes.

Hence, it is still under discussion whether the lifetime of transitional flow structures would diverge or converge with increase in Reynolds number. Moreover, it is not known if there is a threshold Reynolds number at which the lifetime dependency changes. Hence investigations on the lifetime (decay) of transitional flow structures at low Reynolds number are needs to gain a deep insight into the transitional phenomena in pipe flows.

As a part of investigations related to laminar to turbulent transitions in pipe flows at low Reynolds number, the existence of coherent structures, which are called traveling waves, was first reported by Faisst & Eckhardt (2003) and Wedin & Kerswell (2004). Traveling waves were observed experimentally by particle image velocimetry measurements carried out by Hof *et al.* (2004), showing that they were periodically arranged fast streak structures near the pipe wall and slow structures in a puff cross-section. The investigations carried out with experimental and numerical approaches by Hof *et al.* (2005), Schneider *et al.* (2007*a*), Schneider *et al.* (2007*b*), Kerswell & Tutty (2007), Eckhardt *et al.* (2007) and Willis & Kerswell (2008) were aimed at discovering the characteristics of coherent structures in detail. Characterizing transitional flow structures in low Reynolds number pipe flows may answer the questions of e.g. why they are localized in a certain Reynolds number range and how long they would remain in a pipe while convecting downstream. Other than that, the construction of a correct formalism regarding coherent structures in pipe flows would enable the interpretation of dynamic systems through unstable solutions, hence it is an outstanding challenge in fluid mechanics.

1.5 Contents of the thesis

Despite many successful investigations described in the previous sections, there are a large number of issues still to be investigated regarding laminar to turbulent and also turbulent to laminar transitions in pipe flows. The present investigations were concentrated on clarifying those issues, especially regarding the critical Reynolds number of transitional pipe flows, the propagation characteristics of puffs and slugs in pipes and the lifetime of transitional flow structures in low Reynolds number pipe flows.

The contents of the present thesis are as follows. Chapter 1 is the present introduction with a brief literature survey, where issues related to the thesis topics are introduced and the necessary investigations to be studied further are described. Careful experiments were carried out for the investigations by employing a test rig, described in detail in chapter 2, which is a modified and extended version of the test rig developed by Durst & Ünsal (2006). For triggering the flow, ring obstacles and also an iris diaphragm system were employed for precise measurements, as described in chapter 2. The ring obstacle is a thin metal sheet ring which was manufactured carefully by laser cutting. The iris diaphragm system was electrically driven, which creates instantaneous small flow blockages for a preset lapse time, down to several tens of milliseconds. The iris diaphragm triggering technique permitted the repeatable generation of a single puff or a slug at an arbitrary Reynolds number. Hence, in this way, the device greatly facilitated accurate measurements for determining the dynamic characteristics of puffs and slugs. The measurement techniques employed (using a hot wire anemometer and a pressure transducer) are explained in detail also in chapter 2 for better understanding of the results presented in the following chapters.

First, preliminary experiments were carried out to verify that the test rig employed functioned in a well-controlled manner. Accordingly, in chapter 3, it is described that velocity profiles of a pipe for different Reynolds numbers and different pipe lengths is shown and the natural transition occurring at the critical Reynolds number is shown with velocity-time profile measured by hot wire anemometry. Then the laminar to turbulent transitions triggered by ring obstacles are explained. Various heights of ring obstacles were employed and the results are introduced such as the dependence of ring height and flow state, which was triggered by the ring obstacles, on the critical Reynolds number. A set of measurement was carried out also with a large diameter pipe test rig which flow was triggered by ring obstacles to suggest the pipe diameter dependence on transitions nature in pipes.

One of the focal points of the present work was the investigation of the development of puffs and slugs in a pipe from upstream to downstream at different Reynolds numbers in pipe flows, presented in chapter 4. A single flow triggering technique, employing the iris diaphragm system described in chapter 2, was applied to investigate the dynamics of the development of puffs and slugs, from their generation at the pipe inlet over the entire pipe length. Velocity measurements with a hot wire anemometer, located at the exit of the pipe, were performed to detect puffs and slugs and the signals were later processed to obtain ensemble averaged signals. The measurements were conducted with different pipe lengths starting from 0.5 m and extending to 8 m, so that velocity measurements could be carried out at each pipe length to determine the accurate propagation velocities. The development of puffs and slugs and the increase in the propagation velocity with increase in pipe length at different Reynolds number are also shown in chapter 4.

Splitting of puffs, occurring at different locations in a pipe, could be observed in certain Reynolds number range, between a Reynolds number which was slightly higher than that where typically equilibrium puffs were observed and a Reynolds number which was lower than that where typically slugs were observed. Accordingly, through the observation of puff splittings, the transformation from puffs to slugs could be well described in chapter 4.

In chapter 5, investigations on the decay of puffs, which were carried out by direct measurement of 'full-lifetime' of transitional flow structures at low Reynolds number, are reported. The experimental setup employed in the previous investigations was further modified to obtain a higher resolution in controlling Reynolds number. The measurements were carried out with a hot wire anemometer and a pressure transducer. Full-lifetime of transitional flow structures could be measured by a new method by obtaining the pressure difference between two locations in a pipe, one at the inlet (after the triggering location) and the other at the position near the pipe outlet. Through the pressure signal, the development and decay of transitional structures could be clearly shown. Since the transitions occurring at low Reynolds number are statistical phenomenon, it was necessary to repeat a number of measurements to obtain reliable results. The results of full-lifetime measurements for a pipe length $L = 633D$ are presented separately for $Re < Re_{crit}$ and $Re \geq Re_{crit}$ where Re_{crit} was defined as the critical Reynolds number at which the triggered structures start to develop into sustained puffs in chapter 5.

In chapter 6, conclusions summarizing all the results obtained from the different investigations are presented. There it is attempted to characterize laminar to turbulent transitions in pipe flows extensively.

Finally, in chapter 7, as an outlook and a bridge for subsequent investigations, Reynolds stress anisotropy measurements is introduced. Accordingly, the results of turbulent Reynolds stress anisotropy of slugs, measured with an x-probe hot wire anemometer were described.

Chapter 2

Test rig for the investigations

The experimental test rig applied for the present investigations is systematically explained in this chapter, and is shown schematically presented in figure 2.1. The test rig was developed first by Durst & Ünsal (2006) and then modified to conduct accurate measurements in the present investigations. The test rig contains mainly a mass flow rate controller, a flow conditioner, a pipe flow test section, a triggering device, a hot wire anemometer system and a computer operated with a special data acquisition system. Each component of the test rig is explained in the following sections. The measurement techniques applied are also described in this chapter in detail.

2.1 Mass flow rate control system

Two types of a mass flow control unit (MFCU1 and MFCU2) were employed for different investigations described in the present thesis. In chapters 3, 4 and 6, MFCU1 was employed, which was first developed and described in detail by Durst *et al.* (2003). MFCU1 contains mainly a critical valve, an electric valve-positioning controller and a pressure sensor. MFCU1 was connected to an air supply with a pressure of 5 bar. The principle of the function of MFCU1 is briefly explained here, referring figure 2.2 for the sake of completion.

The high-pressure reservoir is filled with air of predetermined pressure P_H, density ρ_H and temperature T_H. From the high-pressure chamber to

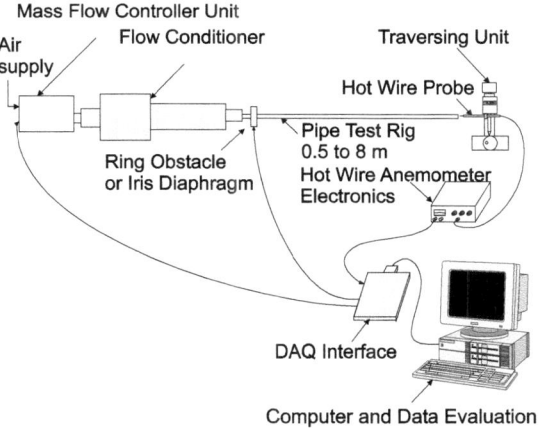

Figure 2.1: Schematic presentation of the experimental test rig, equipped with a ring obstacle or an iris diaphragm system as a triggering device (the triggering device was removed for the investigation of natural transition)

the low-pressure chamber, a flow with velocity U is established through the nozzle when a pressure difference $P_H - P_L$ is imposed. The change of the thermodynamic state at the nozzle with respect to the state in the chamber is described by a steady-state energy equation which, for non-dissipative flows under adiabatic conditions, may be written as

$$U = \sqrt{2 c_p T_H \left(1 - \frac{T}{T_H}\right)} \qquad (2.1)$$

where c_p is gas specific heat at constant pressure.
Rewriting equation 2.1 using $2 c_p T_H = 2\kappa P_H / (\kappa - 1)\rho_H$, where κ is heat capacity ratio c_p/c_v, yields

$$U = \sqrt{\frac{2\kappa P_H}{(\kappa - 1)\rho_H} \left(1 - \frac{P}{P_H}\right)^{\frac{\kappa-1}{\kappa}}} \qquad (2.2)$$

where P the pressure at the nozzle. For the present considerations, U at the lower pressure chamber is of interest, which is referred to as U_L and the corresponding density as ρ_L. Hence the mass flow rate density θ can be written as

$$\theta = \rho_L U_L = \rho_H \left(\frac{P_L}{P_H}\right)^{\frac{1}{\kappa}} \sqrt{\frac{2\kappa P_H}{(\kappa - 1)\rho_H} \left(1 - \frac{P}{P_H}\right)^{\frac{\kappa-1}{\kappa}}} \qquad (2.3)$$

If one employs only a converging nozzle, a continuous decrease in the pressure ratio P_L/P_H results in the critical pressure condition

$$\frac{P^*}{P_H} = \left(\frac{2}{\kappa + 1}\right)^{\frac{\kappa}{\kappa-1}} \qquad (2.4)$$

A further decrease in the pressure $\frac{P_L}{P_H} < \frac{P^*}{P_H}$ does not change the density ρ_L and the velocity U_L. Hence one obtains for sub-critical pressure conditions, i.e. $\frac{P_L}{P_H} < \frac{P^*}{P_H}$,

$$\theta_L = \theta_{max} = \rho^* U^* = \left[\rho_H \left(\frac{2}{\kappa+1}\right)^{\frac{1}{\kappa-1}}\right] \left[U_{max} \sqrt{\frac{\kappa-1}{\kappa+1}}\right] \qquad (2.5)$$

or, by introducing

$$\theta_L = \left[\rho_H \sqrt{\frac{2\kappa P_H}{(\kappa-1)\rho_H}}\right] \left[\sqrt{\frac{\kappa-1}{\kappa+1}} \left(\frac{2}{\kappa+1}\right)^{\frac{1}{\kappa-1}}\right] \qquad (2.6)$$

expressing θ_L in terms of P_H and T_H yields

$$\theta_L = \frac{P_H}{\sqrt{RT_H}} \left[\sqrt{\frac{2\kappa}{\kappa+1}} \left(\frac{2}{\kappa+1}\right)^{\frac{1}{\kappa-1}} \right] \qquad (2.7)$$

where **R** is the gas constant.
Hence the mass flow rate density θ_L depends on the pressure and temperature in the high-pressure chamber (P_H and T_H) and on κ of the gas. However, it does not depend on the pressure and temperature conditions in low-pressure chamber. Finally, one obtains a mass flow rate equation as follows:

$$\dot{M} = \theta_L F = \frac{P_H}{\sqrt{RT_H}} \left[\sqrt{\frac{2\kappa}{\kappa+1}} \left(\frac{2}{\kappa+1}\right)^{\frac{1}{\kappa-1}} \right] F \qquad (2.8)$$

where F is the area of valve opening. The valve opens and closes according to the input signals given by the computer, being regulated by the positioning sensor of the valve rod. When the ratio of P_H and P_L satisfies the condition as described above, defined by a gas dynamic equation, the flow is choked, i.e. independent of the pressure change in the lower pressure chamber and dependent only on the thermodynamic properties of the high-pressure chamber such as pressure P_H and temperature T_H and the valve opening area F. The same principle was applied for the experimental investigations of transitional pipe flows by Rotta (1956), as briefly described in section 1.1. In the present test rig, there was no remarkable change in T_H which affects the deviation in controlling the mass flow rate.

While opening the valve, P_H decreases, which causes a change in the mass flow rate, hence a pressure sensor is also installed on the higher pressure chamber of MFCU1 to compensate for the drop in P_H. As mentioned in section 1.1, it is an important issue for investigations of transitional phenomenon in pipe flows that the mass flow rate is independent of the pressure change of the lower pressure side, since when a laminar to turbulent transition occurs, the pressure of the pipe (P_L, lower pressure side of MFCU1) suddenly drops. An input signal of 0 to 10 V was applied to the MFCU1 so that it controls the mass flow rate from 0 to ca. 3.9 kg/s, which is equivalent in Reynolds number from 0 to ca. 17000 for a $D = 15$ mm pipe, where D is the pipe diameter. The possible minimum control step of MFCU1 in

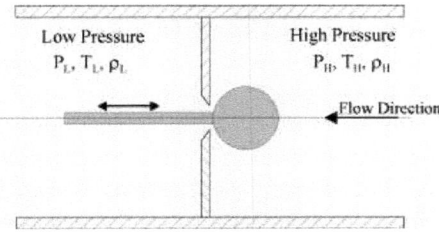

Figure 2.2: Schematic representation of the inside of MFCU1 showing a valve and a high- and a low-pressure reservoir

Reynolds number was 50 to 100 within an error of ±1%. Figure 2.3 shows the calibration curve given by Ünsal (2006).

For the investigations described in chapter 5, another mass flow controller (Bronkhorst, F-202AC-FB-44V) was employed since it had a better resolution in controlling the Reynolds number of the pipe flow. With an input signal from 0 to 10, the mass flow controller could supply a mass flow which is equivalent in Reynolds number from 0 to ca. 5800 for every 5 steps within an error of ±0.4% (cf. figure 2.4). As the mass flow controller did not contain any function of the critical nozzle concept described above, it was necessary to employ an additional critical nozzle to keep the mass flow rate constant and independent of the pressure change in the downstream direction, caused by transition occurrences. The critical nozzle was made of aluminum and was manufactured in the workshop of the Institute of Fluid Mechanics at the Univeristy of Erlangen-Nuremberg (LSTM-Erlangen). The critical nozzle was connected to the outlet of the mass flow controller in order to let the mass flow be choked. The combined unit of the mass flow controller and the critical nozzle is called MFCU2. Before starting preliminary measurements, MFCU2 was calibrated using a hot wire anemometer. The calibration results confirmed that the mass flow rate is linearly correlated to the input signal, as confirmed by data supplied by the manufacturer, shown in figure 2.4. It was thus verified that MFCU2 was insensitive in controlling the mass flow to the pressure change due to the occurrence of downstream transitions, hence

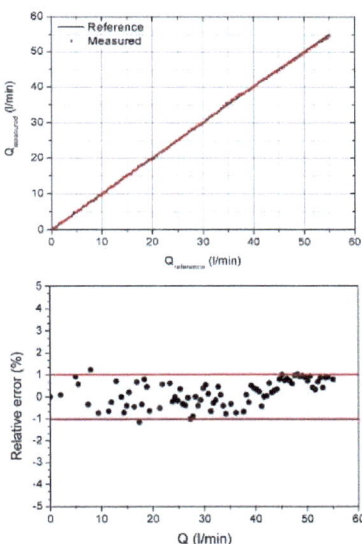

Figure 2.3: Calibration curve for MFCU1 given by Ünsal (2006)

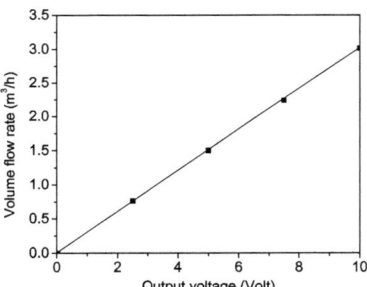

Figure 2.4: Calibration curve according to data supplied by the manufacturer

it was suitable for the present investigations. The resolution in controlling the mass flow rate of MFCU2 was also checked and it was found that it was able to control the flow in each 5 steps in terms of Reynolds number.

The input signal for the above two MFCUs was set and given by a computer through a data acquisition facility with a specially written program.

2.2 Flow conditioner and pipe test rig

Adjacent to the MFCU, a flow conditioner was fitted which consists mainly three parts, a plenum chamber, a honeycomb structure section and a smooth inlet nozzle. In the plenum chamber, acoustic pressure disturbances were suppressed, and in the honeycomb structure, irregularities of flows were restrained. Subsequently, the flow passes through the inlet nozzle to the pipe section. The inlet nozzle is designed such that no flow separation would occur in it. An inlet flow (a flow from the outlet of the flow conditioner) showed a block profile and its condition was free from any large irregularities, such as fluctuations typically found in turbulent flows, as checked by a hot wire anemometry measurement; however, a flow may contain triggering origins because of the form of the velocity profile originating at the inlet nozzle, as pointed out by Durst (2008). Hence further investigations would be necessary to establish the dependence of the inlet conditions, e.g. inlet flow velocity profile, on natural transitions at very high Reynolds numbers. Nevertheless, the present investigations were concentrated on transitions triggered by obstacles which occur much below the critical Reynolds number, hence the inlet flow condition should not be a major issue.

Sustained brass pipes are employed for the pipe section of the experimental test rig. They were manufactured in a seamless manner, because seams might cause unfavorable transitions to be triggered or distort the velocity profile of the flows. The pipe inner diameter was $D = 15$ mm with a high accuracy of finishing, as it deviates by less than 1%. A number of pipes were carefully connected to give various total pipe lengths from $L = 0.5$ to 8 m, being changed occasionally for present investigations. When connecting one pipe to another, special care was taken: the inner edge of the pipe outlet was first cleaned with a sharp knife and then the pipes were connected carefully to each other so that there was no excentricity or discontinuity (especially protuberances) which might cause unfavorable transitions in pipe flows. The connection point of two pipes was covered with a larger inner diameter brass pipe for sealing and also for holding the connected part straight. After the connection of the pipes, their alignment was checked with a water level in a vertical and a horizontal direction so that substantial misalignment was avoided. For each pipe length, laminar velocity profiles were first checked before actual measurements to ensure that the flow was symmetrical and in a good condition, as described in chapter 3. Sustained brass pipes of $D = 40$ mm were also employed for a small part of the present investigations. The length of the $D = 40$ mm pipe was set at $L = 8$ m and the pipes were carefully connected in the same way as for the $D = 15$ mm pipe test rig.

2.3 Devices for triggering laminar to turbulent transitions

At the start of the experimental investigations, natural transitions in which no triggering device was employed, where investigated. Then, for investigations of transitions in pipe flow in more detail, transitions triggered by triggering devices, such as ring obstacles and an iris diaphragm system, were

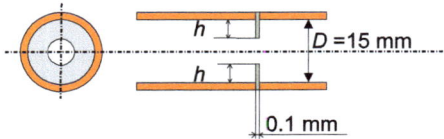

Figure 2.5: Schematic diagram of a ring obstacle geometry

investigated. The ring obstacles, made of a thin metal blade of thickness 0.1 mm, were manufactured by laser cutting, shown schematically in figure 2.5. It was necessary for the ring obstacles to be manufactured by laser cutting to keep the inner edge sharp since it is exposed to the flow and also the ring cross-sectional form should have no discontinuity. The ring obstacles were basically static triggering devices, which shrunk the cross-sectional area suddenly at their location. Consequently, they disturbed the flow with distorted its velocity profile continuously in time. By replacing the ring with a varying ring height h mm, the disturbance amplitude could be varied. It is expected that as h (and also the closing area ratio) increases, the amplitude of the disturbance is also increased. The investigations of transitions triggered by ring obstacles are described in chapter 3, where the disturbance amplitude is represented by the ring height h or the closing area ratio as a percentage based on the cross-sectional area. The location where ring obstacles were fitted was varied from the pipe inlet region, as shown in figure 2.1, to downstream, to demonstrate the dependence of triggered transitions on the triggering locations, where the flow was developing or fully developed, and the results are presented in chapter 3.

Ring obstacles were further employed to investigate transitions in pipe flows also at various Reynolds numbers, which revealed many interesting results; however, the transitions triggered by ring obstacles occurred intermittently and the transitional flow structures were not always a single one but sometimes multiples as a result of combinations with each other, as shown in chapter 3. Therefore, for obtaining a deeper insight into transitional flow structures, it was necessary to apply a triggering device with variable amplitudes which can trigger a single transitional flow structure deterministically.

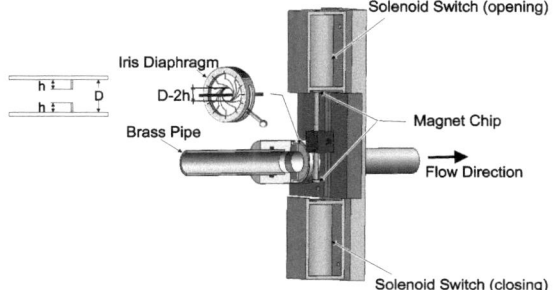

Figure 2.6: Three-dimensioned representation of the iris diaphragm triggering system

For this purpose, an iris diaphragm system was applied, which was first developed by Durst & Ünsal (2006) and is shown in figure 2.6. The iris diaphragm (Melles Griot GmbH) was made of several thin metal blades. The cross-section of an iris diaphragm while closing appears as a circle (axisymmetric) like a ring obstacle, thus a disturbance to a flow was generated when the iris diaphragm was closed and the disturbance was removed on reopening it. A shutter, which functions similarly to an iris diaphragm and has an advantage because it is commercially available with an electrical device to close and reopen it, i.e. favorable for the accurate control of triggering, was not applied in the present investigations because its cross-section while closing was not axisymmetric.

As no close-reopen controller for an iris diaphragm like a shutter was commercially available, a suitable electrical controller for the iris diaphragm was manufactured additionally at the workshop of LSTM-Erlangen. First, a magnet switch was applied by Durst & Ünsal (2006) so that the iris diaphragm closed fast enough, and to reopen it a metal spring was applied. In the present investigation to study a single transitional structure in detail, it was attempted to have very short lapse time (duration of iris diaphragm closing and reopening), so the iris diaphragm was modified by employing another magnet switch. Finally, two magnets were operated by a 25 V electrical current to achieve independently controlled fast closings and reopenings of the iris diaphragm (see figure 2.6). The magnets were operated when a signal was given by the computer through a data acquisition system. The closing height h of the iris diaphragm was adjusted to a predefined value according to the findings of the investigations on transitions triggered by ring obstacles. The lapse time of the iris diaphragm was also controlled electrically through a driving signal imposed on the device by the computer employed. The smallest lapse time that could be realized with the present system was 30 ms, which was short enough to generate a single puff and slug in all ranges of Reynolds number below the critical Reynolds number in the present investigations. Since 30 ms was the shortest lapse time that could be applied in a well-controlled manner, the effect of shorter lapse times could not be investigated. Nevertheless, it was observed that the iris diaphragm did not trigger flows at all if the lapse time was too short, which might be related to a structural issue of flows related to the origin of transitions in pipe flows, which should therefore be investigated. In contrast, if the lapse time exceeded a certain value, the iris diaphragm functioned like a ring-type obstacle, i.e. it disturbed the flow for a longer time and, as a consequence, transitional flow structures appeared as two or more combined which was no longer possible to be investigated as a single transitional flow structure. The issues described above are discussed in chapter 4.

2.4 Measurement equipment and data acquisition system

Throughout all the present experimental investigations, velocity measurements were carried out using a single hot wire probe, connected to DISA 55 M01 constant-temperature hot wire anemometer electronics. A measured signal potential from the anemometer E can be related to the cooling velocity U, which is the velocity perpendicular to the wire and fully correlated over the entire length of the wire, by the following simpled equation (see, e.g., Brunn (1995)):

$$E^2 = a + bU^{0.45} \qquad (2.9)$$

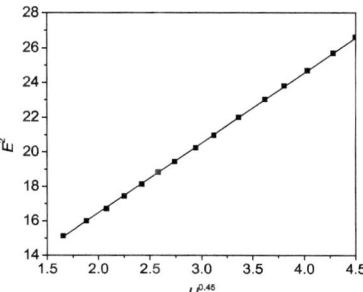

Figure 2.7: Typical calibration curve for hot wire anemometry

This equation shows a linear relation between E^2 and $U^{0.45}$ correlated by parameters a and b which can be obtained by a calibration of the hot wire. An example curve of calibrations of the hot wire to determine a and b is shown in figure 2.7. Calibration of the hot wire anemometer was conducted prior to each set of measurements in order to ensure being accurate in the velocity measurements, employing a calibration nozzle system (Dantec). Since the velocity U is a cooling velocity, a hot wire is sensitive to the temperature change during measurements. Thus it was required to measure the atmospheric temperature by a thermocouple simultaneously in the calibration and also in the actual measurements to minimize errors of measurement caused by temperature deviations. The hot wire was mounted on a traversing unit at the pipe outlet. For measuring velocities at different positions in the cross-section of the pipe, the vertical motion of the traversing system was activated by the computer to change the wire position. The output signal of the hot wire anemometer, temperature and positioning sensor were simultaneously measured and recorded by a computer through a 16-channel, 16-bit, 333 kHz data acquisition card. The flow rate of the MFCU and the position of the hot wire were set by the computer. Several special software programs, together with various sub-programs, were written for the data acquisitions described above to ensure that all measurements were conducted in a well-controlled manner.

For measurements of the full-lifetime of transitional flow structures described in chapter 5, a pressure transducer (RCI, PD 2915 SM1) was employed. Pressure measurement holes were bored in the pipe wall directly with a hole diameter of $\phi = 0.5$ mm. After boring the holes, the pipe inner surface was properly cleaned to avoid any discontinuity which might cause unfavorable transitions or velocity profile distortions. The pressure signal was acquired and recorded by the computer for analysis.

Chapter 3

Preliminary research work

In this chapter, preliminary research work carried out prior to detailed investigations of puffs and slugs is described. First, several verification experiments were performed to show that the test rig employed, introduced in chapter 2, functioned accurately and also in a well-controlled manner. Experimental investigations were conducted by employing a hot wire anemometer (HWA) for flow velocity measurements at the pipe outlet. First the development of a laminar flow is shown in section 3.1, employing pipes of several lengths, i.e. with various distances for flows to develop. The results confirmed that the flow was symmetrical at any location, which was a necessary condition for later investigations. When one keeps on increasing the Reynolds number of a pipe flow, a transition occurs naturally above its critical Reynolds number, originating at the pipe inlet. Reynolds number was defined as $Re = u_{bulk}D/\nu$, where u_{bulk} is the bulk velocity, D is the pipe diameter and ν is the kinematic viscosity of air. The natural transition was observed at an $L = 8$ m pipe outlet by HWA measurement, as shown in section 3.1.

Transitions in pipe flows can be induced by employing triggering devices, e.g. ring obstacles, at a Reynolds number that is lower than the critical Reynolds number. A ring obstacle, made of a thin metal blade, creates a sudden flow blockage at its locus, hence a flow can be triggered to be a transitional flow. Since the blockage area, determined by a ring height h, could easily varied by replacing rings of various heights, first ring obstacles were chosen to investigate the triggered transitional flow phenomenon in pipes. The HWA was set first on the center line, then moved to various positions in the pipe cross-section to yield the velocity-time profile in transitions of pipe flows. The appearance of transitional flows triggered by ring obstacles was found to be similar to the natural transition, i.e. they occurred randomly, hence the timing of occurrence was not under control. However, the phenomena could be comprehensively studied by applying ensemble averaging within the measuring time for statistical evaluations, as shown in section 3.2. Various heights of rings were then applied to show the dependence of the ring height on the transitional phenomena. The locus of ring obstacles was altered from the pipe inlet to the downstream position, where the flow was fully developed, to study about the sensitivity of the flow condition, i.e. developing or fully developed, to be triggered. The results are shown in section 3.3.

Finally, $D = 40$ mm pipe flow transitions, triggered by ring obstacles, were investigated, where a possible dependence of pipe diameter on the critical Reynolds number was found, and compared with several results from the literature as described in section 3.4. In section 3.5, the results obtained in the preliminary research work described in the present chapter are summarized.

3.1 Development of laminar flow and natural transitions

Laminar flow development was first checked by employing different pipe lengths, yielding detailed cross-sectional velocity profile information, as shown in figure 3.1. The velocity profile was measured and is shown in figure 3.1, where the radius r and the time-averaged velocity $\bar{u}$ are normalized by dividing by the radius of the pipe R and by the bulk velocity u_{bulk}, respectively. The development length was calculated using equation 3.1 for $Re = 2855$ and 8575, being ca. $L = 162D$ and $485D$ (2.43 and 7.3 m), respectively. It can be seen that the flow is fully developed only for $L = 8$ m and not for $L = 0.5$ and 2 m at both Reynolds numbers. It is also clear from figure 3.1 that

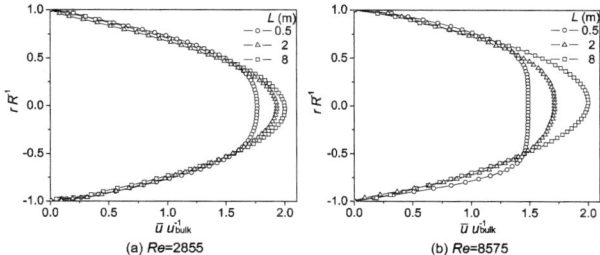

Figure 3.1: Time-averaged velocity ($\bar{u}$) profile at the pipe outlet measured by an HWA for a pipe cross-section at (a) $Re = 2855$ and (b) $Re = 8575$ for $L = 0.5$, 2 and 8 m pipes, presented in a normalized manner as radial position r is divided by the pipe radius R and $\bar{u}$ is divided by the bulk velocity u_{bulk}

the flow is axisymmetric, which verifies that the further investigations were carried out with well-conditioned flows. The results are in good agreement with corresponding flow predictions of the development length correlation proposed by Durst *et al.* (2005):

$$\frac{L}{D} = \left[(0.619)^{1.6} + (0.0567 Re)^{1.6}\right]^{\frac{1}{1.6}} \qquad (3.1)$$

where L is the pipe length and D is the pipe diameter.

Hence for every investigated flow at any Reynolds number, the development length required to yield a parabolic velocity profile could be calculated with equation 3.1. This information permitted the state of the flow to be assessed for the investigated pipe lengths, i.e. the state that the flow would have without any flow disturbances.

After checking the condition of the flow, the first investigation regarding transitions in pipe flows was conducted. A naturally occurring transition was observed for the test rig with a pipe length $L = 8$ m. For this investigation, the test rig introduced in chapter 2 without any triggering device was employed. The mass flow rate was set and maintained constant while the velocity of the flow was measured by the HWA at the center line of the pipe outlet for many minutes, while the HWA signals were recorded continuously

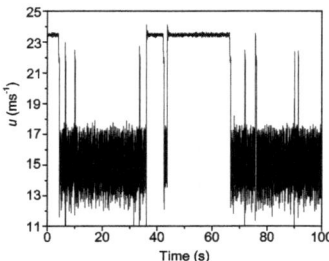

Figure 3.2: Velocity-time profile measured by the HWA on the center line of the $L = 8$ m pipe outlet at around the critical Reynolds number, $Re \approx 11500$

by a computer. On increasing the mass flow rate controller input gradually, the Reynolds number of the pipe flow was also increased and then a first transition appeared, which could be detected from the HWA records. Figure 3.2 shows a typical HWA velocity-time profile for a naturally occurring transitions at around the critical Reynolds number of the system, $Re_{crit} \approx 11500$. The critical Reynolds number was defined as the Reynolds number at which a first transitional structure starts to appear at the $L = 8$ m pipe outlet. Reynolds (1883) showed that a laminar to turbulent transition in pipe flows would occur intermittently (cf. figure 1.2) and such characteristics are also shown in figure 3.2.

When a transition occurs in a pipe, the velocity at the center line drops suddenly according to the change in the pipe cross-sectional velocity profile from laminar to turbulent, as shown schematically in figure 3.3. When the transition occurs the other way round, from turbulent to laminar, the velocity increases back to the laminar flow value. A single portion of such transitional structures, such as shown in figure 3.4, is called a slug, which generally appears at $Re \geq 3000$ in pipe flows. Figure 3.4 illustrates typical features of slugs already pointed out by Wygnanski & Champagne (1973) (cf. section 1.3).

Since natural transitions occur non-deterministically, one should apply ensemble averaging to analyze them so that the transitional phenomena could

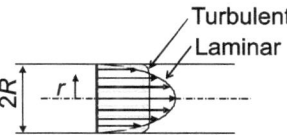

Figure 3.3: Schematic diagram of a laminar and a turbulent velocity profile, showing their crossing point

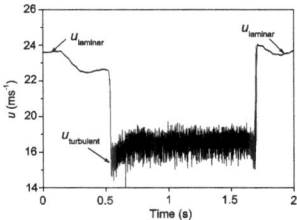

Figure 3.4: A single portion of a natural transition structure (a slug) measured by the HWA on the center line of the $L = 8$ m pipe outlet at $Re = 11500$

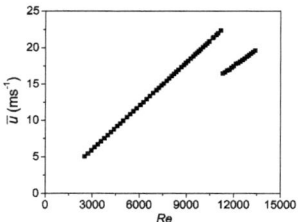

Figure 3.5: Time-averaged velocity $\bar{u}$ on the center line at the $L = 8$ m pipe outlet measured by the HWA with increasing Reynolds number

be quantified, hence the time-averaged velocity $\bar{u}$ was obtained using the following equation:

$$\bar{u} = \frac{\sum_{n=1}^{N} u}{N} \qquad (3.2)$$

where u is the instantaneous velocity and n and N are the ordinal and total number of samplings, respectively. The variation of $\bar{u}$ with increasing Reynolds number for natural transitions is illustrated in figure 3.5, which shows that as the Reynolds number increases, $\bar{u}$ first increases but, once a transition starts to appear, it drops suddenly and then starts to increase again. Hence it can be readily recognized from the results such as those in figure 3.5 what the critical Reynolds number of the system is under the present condition. For further analysis, the turbulent intensity Tu is plotted in figure 3.6. $Tu = u'/\bar{u}$, where $\bar{u}$ is a time-averaged velocity as defined in equation 3.2 and u' is the root mean square (rms) of the velocity, defined as

$$u' = \sqrt{\frac{\sum_{n=1}^{N}(u - \bar{u})^2}{N}} \qquad (3.3)$$

where n, N and u are the same as in equation 3.2.

With increasing Reynolds number, both u' and Tu suddenly jump to a higher value after the natural transition starts, also indicating apparently the critical Reynolds number of the system.

The above-mentioned results clearly showed that the test rig permitted the pipe flow to remain laminar up to $Re_{crit} \approx 11500$ over a length of $L = 8$ m, corresponding to $L/D = 533.3$, with a pipe diameter $D = 15$ mm.

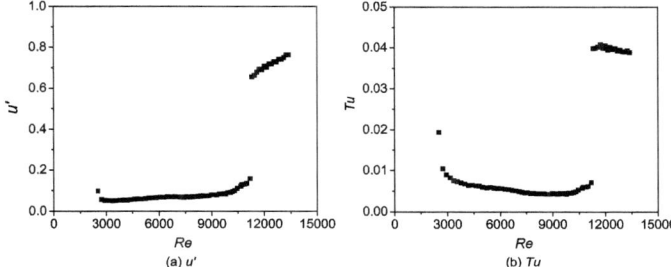

Figure 3.6: Changes in the values of (a) u' and (b) Tu based on velocity measurements carried out with the HWA on the center line of the $L = 8$ m pipe outlet with increase in Reynolds number

An interesting observation during the preliminary experimental work is mentioned here. A kind of transitional structure, which looked like a turbulent spot, was detected by the HWA at the $L = 0.5$ m pipe outlet at around the critical Reynolds number, the velocity signal of which is shown in figure 3.7. This result implied that such a short-term transitional flow structure developed into a slug while propagating downstream in the pipe. The finding gave the motivation to study the development of the transitional flow structures in the pipe, and the results of that investigation are presented in chapter 4.

3.2 Transitions triggered by ring obstacles

As the second part of preliminary research investigations, ring obstacles were employed to study the triggered laminar to turbulent transition in pipes. The ring obstacles were made of a thin metal sheet, of thickness 0.1 mm, and were manufactured using laser cutting. The geometry of the ring obstacles employed is shown schematically in figure 2.5. The ring obstacle was fitted to the pipe inlet region at the locus shown in figure 2.1. By replacing rings of various height h, the amplitudes of disturbances to the flows could be

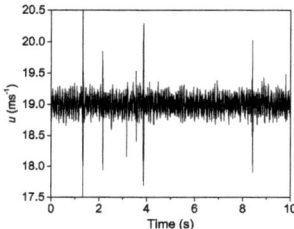

Figure 3.7: Transitional flow structures detected by the HWA at the $L = 0.5$ m pipe outlet at around the critical Reynolds number, $Re \approx 11500$

varied. A systematic relation was obtained between the ring height h and the Reynolds numbers, at which a first transition appeared, triggered by the ring obstacle as shown in the following. The ring height h was varied from 0 to 2.2 mm, which corresponds to a 0 to 50% closing area ratio based on the cross-section of the pipe.

First, an $h = 2.2$ mm (corresponding to a closing area ratio of 50%) ring obstacle was applied to study the characterics of triggered transitions. The ring was adjusted at the inlet of the pipe and then Reynolds number was increased gradually, while the velocity at the pipe outlet on the center line was measured to detect the transitions. At each Reynolds number, the velocity signal was recorded for some minutes to find where the first transition occurred. The velocity u showed similar characteristics to those of the natural transition as it occurred occasionally and intermittently, as illustrated in figure 3.8, which shows the center line velocity measured by the HWA at three different Reynolds numbers. As the Reynolds number increases, the appearances of transitional structures are more frequent and sometimes they combine with each other. Rotta (1956) introduced an intermittency factor γ, the ratio of the summation of turbulent time t_{turb} and total measurement time T, defined as follows, to describe quantitatively the transitional phenomena occurring occasionally (see also section 1.1):

$$\gamma = \frac{\Sigma t_{turb}}{T} \qquad (3.4)$$

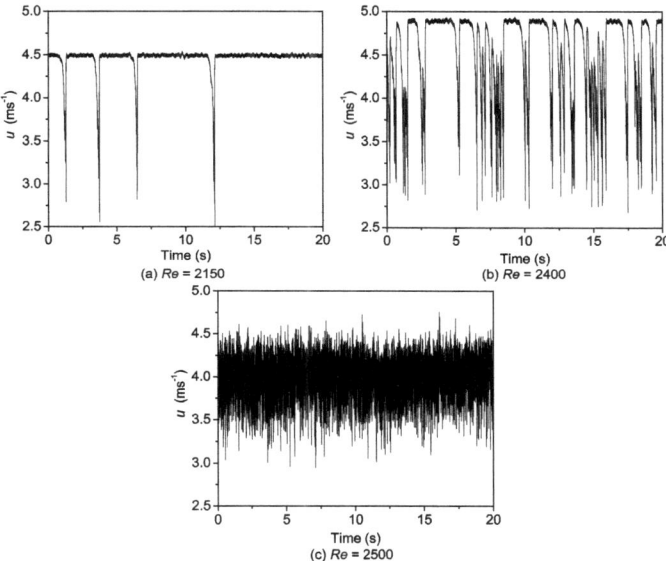

Figure 3.8: Velocity-time profile with a flow triggered by an $h = 2.2$ mm ring obstacle at (a) $Re = 2150$, (b) $Re = 2400$ and (c) $Re = 2500$

where $\gamma = 0$ and 1 indicate that the flow is continuously laminar and turbulent, respectively.

The intermittency factor γ is plotted for the transition triggered by an $h = 2.2$ mm ring obstacle with increasing Reynolds number in figure 3.9, where parts (a), (b) and (c) correspond to the three cases (a), (b) and (c) in figure 3.8. The Reynolds number at which the intermittency curve starts to increase indicates the Reynolds number at which first transitions were induced for the given disturbance. It is interesting that the change in γ from 0 to 1 extends over a Reynolds number range of ca. 400 for the transition at around $Re \approx 2000$ but ca. 100 at around $Re \approx 11500$, i.e. at higher Reynolds number the transition is completed faster than at lower Reynolds number.

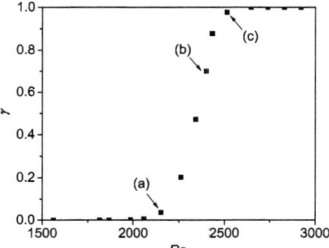

Figure 3.9: Intermittency factor γ of the transition triggered by $h = 2.2$ mm ring obstacles for $1500 \leq Re \leq 3000$ with increase in Reynolds number; $\gamma = 0$ and 1 indicate that the flow is continuously laminar and turbulent, respectively

As in the investigations of natural transitions, the ensemble-averaged center line velocities $\bar{u}$ and u' were obtained for a quantitative description at each Reynolds number of the transition triggered by an $h = 2.2$ mm ring obstacle, as shown in figure 3.10. It is shown that $\bar{u}$ increases with increase in Reynolds number and, after the transition starts to appear, i.e. as γ starts to have a value larger than 0, it decreases and then $\bar{u}$ starts to increase again, when γ reaches 1 (figure 3.9). In figure 3.10(b), u' jumps to a higher value and reaches a peak, then settles, correlated with the $\bar{u}$ and γ change.

So far the velocity was measured only on the center line of the pipe outlet to show the characteristics of the transitions triggered by a ring obstacle. Then the velocity at different radial positions was also measured for detailed quantification of the phenomenon, using $\bar{u}$ and u' mentioned above. The results are shown in figure 3.11. At the position where $r \approx 5$ mm (corresponding to the position 0.66 times R), the laminar and turbulent velocity profiles cross each other as shown in figure 3.3. At such a position with $r \approx 0.66R$, the velocity value does not change whether the flow state is laminar or turbulent, hence $\bar{u}$ increases continuously with increase in Reynolds number, irrespective of Reynolds number (figure 3.11(a), $r = 5$ mm). The u' values for different radial positions are shown in figure 3.11(b). Most of them

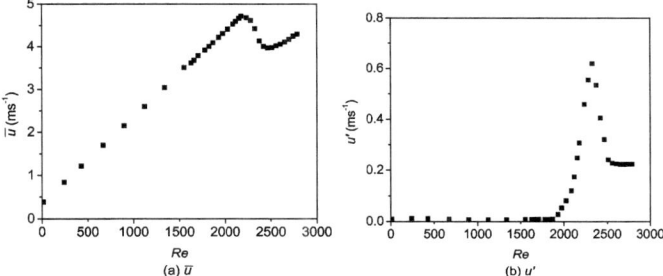

Figure 3.10: (a) Time-averaged velocity $\bar{u}$ and (b) u' value of the transition, triggered by an $h = 2.2$ mm ring obstacle, measured on the center line of the $L = 8$ m pipe outlet measured by the HWA with increase in Reynolds number

show overshooting, but u' for $r = 5$ mm does not because there is no change in the velocity irrespective of the flow state, laminar or turbulent, as is clear from the definition of u' in equation 3.3. Consequently, the curve of u' at $r = 5$ mm is almost identical with the intermittency factor curve, shown in figure 3.9.

In subsequent investigations, rings of various height h were employed to find the relation between h and the Reynolds number at which a transition can be induced by the ring. For convenience, the Reynolds number at which the first transition was observed is termed critical Reynolds number in the following, although the critical Reynolds number of the entire test rig was fixed at $Re \approx 11500$. The investigation procedure was the same as in the previous investigations, first fitting a ring of height h at the pipe inlet, increasing the Reynolds number gradually, measuring the velocity for some minutes at the pipe outlet to observe the appearance of first transitions and determine the critical Reynolds number with the ring obstacle employed. In this investigation, rings of heights $h = 0$ to 1.2 mm (0 to 30% closing area ratio) were employed and the results are shown in figure 3.12. As h increases, the critical Reynolds number decreases, which can be recognized by the jump in the u' value. Hence it is clear that the test rig employed had the possibility

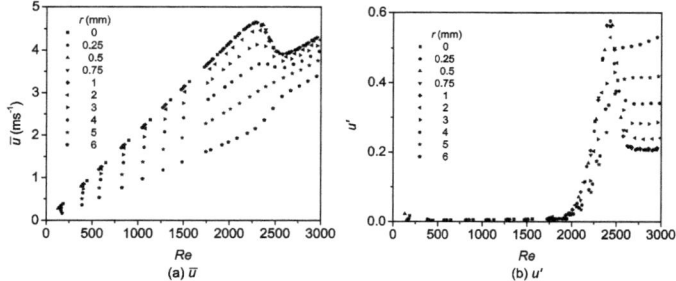

Figure 3.11: (a) Time-averaged velocity $\bar{u}$ and (b) u' value at the $L = 8$ m pipe outlet measured by the HWA for different radial positions with increase in Reynolds number

of triggering flows at arbitrary Reynolds numbers below the critical Reynolds number ($Re \leq 11500$) with variable height h of ring obstacles.

The results shown in figure 3.12 are presented in a different manner in figure 3.13, in which the results obtained by Durst & Ünsal (2006) are also plotted for comparison. Figure 3.13 provides information on the required triggering height h as a function of critical Reynolds number to yield the laminar to turbulent transition of pipe flows. In figure 3.13, (a), (b) and (c) on the line $h = 2.2$ mm correspond to those in figure 3.8. As shown in figure 3.8, when a ring obstacle of a certain height h is adjusted and the Reynolds number is increased gradually, the flow observed at the pipe outlet appears first laminar, then intermittent and turbulent, which can be quantitatively shown by the change in γ from 0 to 1 with increase in Reynolds number. The beginning and end of the transition with increase in Reynolds number are shown with open and filled symbols, respectively, in figure 3.13.

There is a linear relation between the Reynolds number of the first transition and the ring height h for the higher Reynolds number range, but a strong nonlinearity at $Re \approx 2000$ is observed as reported by e.g. Wygnanski & Champagne (1973) and Darbyshire & Mullin (1995). It is well known that at very low Reynolds number, transitions cannot be sustained in

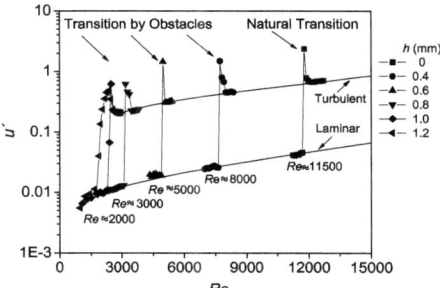

Figure 3.12: Change in u' with increase in Reynolds number for a natural transition and transitions by ring obstacles, obtained by the HWA velocity measurement on the center line of the $L = 8$ m pipe outlet

a pipe, hence with increase in h it is not possible to induce transitions in pipe flows below the minimum critical Reynolds number. The minimum critical Reynolds number, below which no transitional structure can be sustained, will be discussed in detail in chapter 5. The results in figure 3.13 contain the interesting information that the critical Reynolds number even increases with increase in h for $h \geq 1.2$ mm. This phenomenon is due to the flow structures originating at the pipe inlet by the different h values of the rings. This implies that there is a critical height of ring obstacles to trigger the flow at the lowest Reynolds number. Such nonlinearity in the relation between the amplitude of the obstacles (in the present investigation, the ring height h) and the critical Reynolds number at lower Reynolds numbers was also observed in further investigations with larger diameter ($D = 40$ mm) pipe flows triggered by ring obstacles and also $D = 15$ mm pipe flows triggered by an iris diaphragm with different amplitudes. Accordingly, those issues are discussed in later sections.

The flow structures occurring in laminar to turbulent transitions at $Re \lesssim 2300$ have a different nature to slugs and are called puffs. Typical velocity-time profiles at the center line of the pipe measured by the HWA for two types of transitional flow structures, a puff and a slug, are shown in figure 3.14.

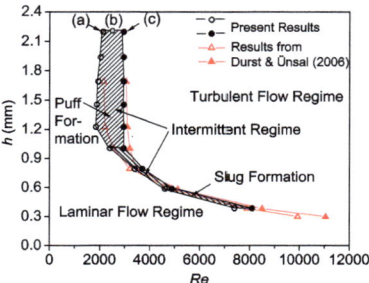

Figure 3.13: Dependence of flow states on Reynolds number and the ring height h

In contrast to slugs, puffs do not show any clear boundary at their front edge (laminar to turbulent transition, early in time), but do at the back edge (turbulent to laminar transition, late in time). Another clear difference between puffs and slugs is the turbulent intensity: the turbulent intensity Tu of a puff is far from fully developed turbulence, unlike a slug. The turbulent intensities of transitional structures were measured and are given in chapter 4. It is also marked on figure 3.13 which type of transitional structures could be observed in which area in the diagram. A similar diagram was drawn by Wygnanski & Champagne (1973), shown in figure 1.4, in which the regions were indicated where puffs and slugs could be observed and between them the area was denoted 'uncertain'. In this range of Reynolds number, some split puffs were observed, the structures of which are neither puffs nor slugs. Hence puff splitting was investigated and is discussed more in detail in chapter 4.

3.3 Triggering of fully developed laminar flow

All the investigations described above were aimed at a better understanding of the laminar to turbulent transition of pipe flows triggered by ring obstacles mounted at the inlet of the pipe. However, the velocity profiles at the pipe inlet were dependent on the inlet flow conditions, which were determined by,

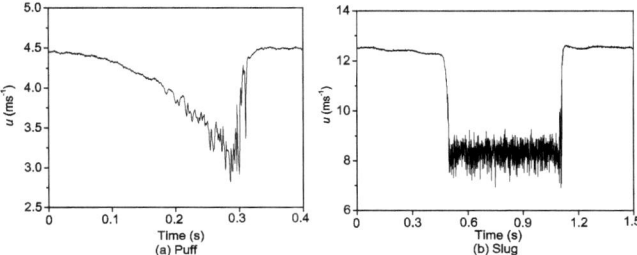

Figure 3.14: Typical velocity-time profile of (a) a puff and (b) a slug, measured at the center line of the $L = 8$ m pipe outlet

e.g., the flow conditioner containing a plenum chamber, honeycomb structures and a nozzle-type flow inlet section to the pipe. The entire inlet flow arrangement also influenced the local wall shear stress at the location where the flows were triggered. Therefore, the flow properties at the location of triggering were not well known and it was decided to carry out an additional set of experiments for which the flow details were well known at the location of triggering.

The test rig in figure 2.1 was employed and an additional nine ring obstacles were manufactured with high precision. The location of the ring obstacles was set at pipe lengths of $L = 5$ and 8 m, at which the flows reached fully developed laminar states up to $Re \approx 5900$ and 9400, respectively, calculated by equation 3.1. Hence, in most sets of experiments described in this section, the flow conditions at the triggering location were the same. Thereafter, an $L = 2$ m pipe was added to ensure the development of triggered flows into puffs or slugs, as ca. $100D$ is the necessary evolution distance mentioned by e.g. Mullin & Peixinho (2006). Then, for each ring obstacle, the mass flow rate was gradually increased until the the first transition was detected at the end of the pipe. The HWA was equipped to detect the first puff or slug as an indication of the critical Reynolds number.

First, the results of flows triggered at $L = 5$ and 8 m are compared with those at the inlet region. The plots in figure 3.15 depict the critical

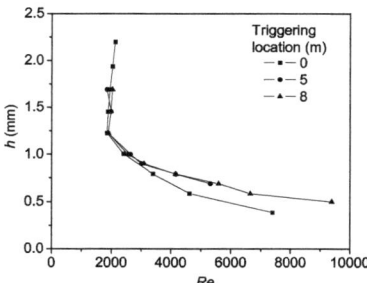

Figure 3.15: Critical Reynolds number comparisons for the triggering locations of ring obstacles

Reynolds number where the first transitional structure appeared with certain ring obstacles with height h. With the same ring of height h, fully developed flow triggered at $L = 5$ and 8 m shows a higher critical Reynolds number than that at $L = 0$ m, hence the fully developed flow was more stable than the inlet flows. The results for triggering locations of $L = 5$ and 8 m merge into each other, which indicates that once the flow is fully developed, the relation between the critical Reynolds number and the ring height h remains the same for any triggering location applied.

The relation between the critical Reynolds numbers and normalized wall fence height (h/D) for pipe lengths of $L = 5$ and 8 m are shown in figure 3.16, in which C indicates a constant value. The results reflect the following:

- According to Jovanović & Pashtrapanska (2004), who examined the laminar to turbulent transition by Reynolds stress anisotropy analysis, the following relation holds:

$$(Re_\lambda)_{crit} = 10\sqrt{5} \quad \sim \quad (Re_D)^{\frac{1}{2}}_{crit} \quad \sim \quad \frac{D}{\lambda} \quad \sim \quad \frac{D}{h} \qquad (3.5)$$

where λ and Re_λ are the Taylor micro scale and the Reynolds number based on the intensity and length scale of the disturbances, respectively.

This readily suggests that

$$\frac{h}{D} \sim \frac{1}{(Re_D)_{crit}^{\frac{1}{2}}} \qquad (3.6)$$

- The experimental findings and the $Re_{crit}^{-\frac{1}{2}}$ relation are plotted in figure 3.16. According to the present results, the exponential factor is -0.56, which is slightly lower than -0.5. Such a small discrepancy between the predicted h/D and Re_{crit} relationships and the experimental findings is due to the remaining experimental scatter of one data set.

The minimum critical Reynolds number at which transitions could be induced by ring obstacles and can be sustained up to $L = 8$ m is $Re_{min} = 1940$, as the vertical line in figure 3.16 shows. This value of the lower critical Reynolds number is very close to the Re_{min} value deduced by Jovanović & Pashtrapanska (2004). They investigated theoretically the minimum critical Reynolds number achievable by considering balances between production and dissipation terms in the turbulence dissipation equation and for pipe flows they obtained $Re_{crit} = 1930$. The transitions at very low Reynolds number in pipe flows are discussed in more detail in chapter 5.

From the results in figure 3.16, it can readily be deduced that

$$h^+ = \frac{h \cdot u_\tau}{\nu} = \frac{h}{D}\sqrt{8Re} \quad \text{since} \quad u_\tau = \sqrt{\frac{c_f}{2}}\bar{u}. \qquad (3.7)$$

Taking into account that $h/D \sim (Re)^{-\frac{1}{2}}$ suggests that $h^+ = $ constant, i.e. it is independent of Re_τ. This is shown in figure 3.17 to be in good agreement with the present experimental findings.

3.4 Large pipe test rig transitions triggered by ring obstacles

For a set of experiments on the pipe with larger diameter $D = 40$ mm, the test rig shown in figure 2.1 was reconstructed. Two MFCU1 units were employed to realize targeted Reynolds numbers. They had the task of supplying

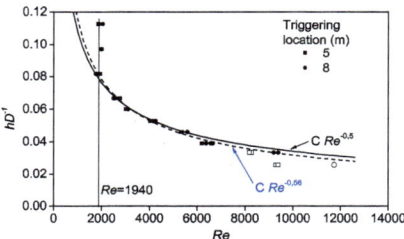

Figure 3.16: Critical Reynolds number dependence on ring height normalized with the pipe diameter D as h/D, for which the flow triggering locations are $L = 5$ and 8 m

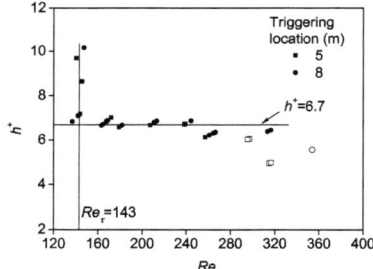

Figure 3.17: Dependence of non-dimensionalized ring height h^+ on a non-dimensionalized Reynolds number Re_τ for fully developed pipe flow at triggering locations $L = 5$ and 8 m. Open symbols are the results of the same experiments in which triggered flows were not fully developed

constant mass flow rates for each set of Reynolds numbers with an accuracy of ±2% independent of any downstream flow condition. This allowed different Reynolds numbers to be set and maintained for each of the attempted investigations. A flow conditioner was designed and applied with a geometry similar to that employed for the $D = 15$ mm pipe. The ring obstacles were precisely manufactured from a thin metal blade as in the previous investigations, and installed at the inlet region.

The critical Reynolds number for the whole setup was higher than $Re = 12000$, which is considerably higher than that for the $D = 15$ mm pipe. Then, the relation between the amplitude of the ring obstacles and the critical Reynolds number was obtained as shown in figure 3.18, where the results for the $D = 15$ mm pipe are also plotted for comparison. For convenience, the amplitude is represented by a closing area ratio in figure 3.18. As for the $D = 15$ mm pipe, the critical Reynolds number decreases linearly with increasing closing area ratio up to $Re \approx 2000$. For $Re \lesssim 2000$, the relation appears nonlinear. According to the results in figure 3.18, a 25% closing area ratio achieves the minimum critical Reanolds number for the $D = 40$ mm pipe, and a 30% ratio for the $D = 15$ mm pipe. This suggests that there should be an analogy between the ring obstacle geometry determined by the height h and the achievable minimum critical Reynolds number, which is not yet well explained. Hence more investigations should be carried out in the future to look into the details of the flow structures after triggering.

Similarity theory suggests that the height necessary to trigger fully developed laminar flow, to cause its transition to turbulence depends on the pipe diameter (see equation 3.6). This is readily suggested from results available in the literature (Barnes & Coker (1905), Darbyshire & Mullin (1995), Draad *et al.* (1998), Hof *et al.* (2003), Lindgren (1969), Meseth (1974), Pfenninger (1961), Reynolds (1883), Rotta (1956), Schiller (1934), Wygnanski & Champagne (1973), Zanoun (2007)), which are plotted in figure 3.19. Hence the existing experimental results suggest that the critical Reynolds number for the laminar to turbulent transition in pipe flow is dependent on the pipe diameter. The larger the pipe diameter, the larger is the critical Reynolds number, as mentioned briefly over 100 years ago by Barnes & Coker (1905). They employed pipes with different diameters from $D = 10.5$ to 54.1 mm.

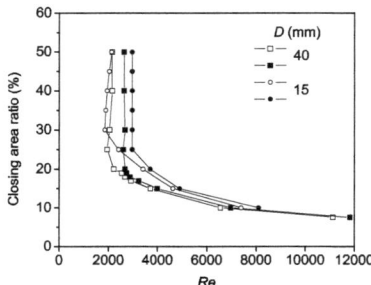

Figure 3.18: Dependence of critical Reynolds number on the closing area ratio for $D = 15$ and 40 mm pipes

For a pipe of diameter $D = 54.1$ mm, they could not obtain a laminar to turbulent transition even at the highest Reynolds number which their test rig could provide, $Re \approx 48000$, whereas for a $D = 19.5$ mm pipe, the critical Reynolds number was $Re \approx 20000$. To investigate these issues further, deduced from the available literature, in the present study the test rig with a larger diameter $D = 40$ mm pipe was employed. For the present test rig, a critical Reynolds number was obtained that was much higher than that obtained for the pipe of $D = 15$ mm, as mentioned before. All this clearly shows that the laminar to turbulent transition of pipe flows is far from being understood, hence further investigations are required.

3.5 Summary of natural and ring obstacle-triggered transitions

Through the preliminary research work described in this chapter, it was first verified that the present test rig was well suited for investigating the laminar to turbulent transitions in pipe flows. The natural transition that occurred in the pipe was studied based on HWA measurements results. Transitions triggered by ring obstacles were then studied and the results were treated by ensemble averaging to evaluate the phenomena quantitatively. Several

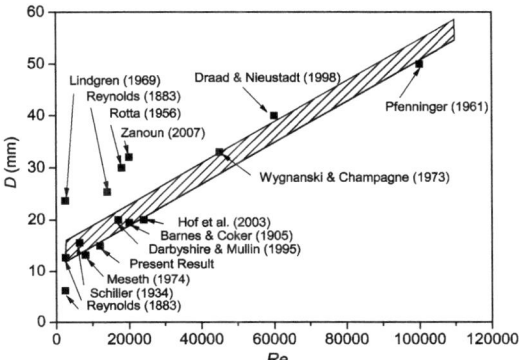

Figure 3.19: Dependence of pipe diameter D on the critical Reynolds number deduced from available experiments

new findings in the investigations were found, e.g. the strange tendency in the relation between ring height h and the critical Reynolds number at $Re \approx 2000$. The dependence of the flow state, developing or fully developed flow, to be triggered was shown by the results for ring obstacles fitted at different positions in pipe. The results apparently show that a fully developed flow is more stable than a developing flow to be triggered when the same height of ring obstacles applied. The relation between the minimum ring height and the critical Reynolds number was investigated and analyzed. The findings suggest the following:

1. The wall fence obstacle height, normalized with the pipe diameter, to trigger the laminar to turbulent transition of a pipe flow changes with Reynolds number as $\sim (Re)_{crit}^{-\frac{1}{2}}$.

2. On plotting the experimental findings as $h^+(Re_\tau)$, it is observed that $h^+ = $ constant. The amplitude perturbation, h^+, is independent of Re_τ. This can be deduced by similarity considerations.

Finally, from the investigations employing the $D = 40$ mm pipe, the possibility of the dependence of pipe diameter on the transitions in pipes is found.

Chapter 4

Development of puffs and slugs

In chapter 3, the natural transition and the transitions triggered by ring obstacles were investigated as preliminary research work. There were a number of interesting findings in those investigations, but the transitions occurred occasionally and intermittently and, moreover, from time to time some structures appeared to be combined with each other as they could be analyzed only by statistical means. Hence for a detailed study of the development of a puff and a slug in a pipe, it was necessary to employ a type of disturbance that could trigger flows to produce a single transitional flow structure. For this purpose, a triggering system with an iris diaphragm, which was first constructed by Durst & Ünsal (2006), was extensively applied after some modifications, as described in chapter 2. The iris diaphragm system permitted the triggering of flows with a predefined amplitude in a well-controlled manner to obtain a single puff and slug at an arbitrary Reynolds number for $Re \leq Re_{crit}$. The closing area and the lapse time of the iris diaphragm system were adjustable for the investigations, which are presented and discussed in the present chapter. From this chapter onwards, experimental investigations were carried out by employing only the iris diaphragm as a triggering device system because it allowed the flow to be triggered deterministically and repeatably with a predefined amplitude. Performing HWA measurements at the pipe outlet of different lengths showed that slugs form and develop basically in the same way at any Reynolds numbers for the high Reynolds number range, as shown in section 4.1, where the dependence of pipe length on the propagation of slugs is also described. On the other hand, puffs form and develop in a pipe in a manner strongly dependent on Reynolds number, as shown in section 4.2, where the dependence of the pipe length on the propagation of puffs is described similarly to investigations of slugs. When studying the development of puffs, puff splitting was observed depending on Reynolds number and also the pipe length. Accordingly, it could be described how the flow structures change from puffs to slugs through puff splitting, discussed in section 4.3, where typical characteristics of split puffs are introduced and the systematic change of the number of splittings with increase in Reynolds number is shown. In section 4.4, the results obtained for the investigation of the development of puffs and slugs in pipe and also the structural change from puffs to slugs are summarized.

4.1 Slug development in a pipe

The development of slugs was first investigated for various Reynolds numbers. The iris diaphragm was set at the pipe inlet. The lapse time of the iris diaphragm system was set at 30 ms, which was the shortest time achievable to close and reopen the present iris diaphragm. Each operation of the iris diaphragm system with a predetermined height of h mm produces a sudden blockage in the flow during operation. By referring to the results discussed in section 3.1 for the dependence of the ring obstacle height on the critical Reynolds number and also on the transitional flow structure form (a puff or a slug), an iris diaphragm closing height h was chosen for the present investigation so as to ensure that it triggered flows to form a slug.

Examples of HWA velocity-time profiles on the center line at the pipe outlet, with pipe lengths of $L = 0.5$, 3 and 8 m, are shown in figures 4.1 - 4.3 at $Re = 4160$, 6250 and 8230, respectively. The time 0 s in the figures indicates the time when the iris diaphragm system was operated, thus all flows in the figures were laminar at time 0 s. As is briefly described in section 3.1, when a slug reaches the pipe outlet, the velocity signal on the center line suddenly drops. The velocity increases again to the laminar value when the slug leaves the pipe completely. A set of time profile measurements is called a realization in the present investigation, and the time between one

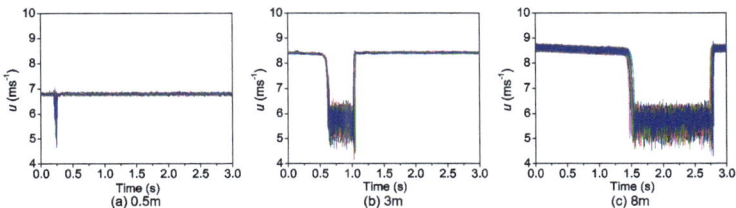

Figure 4.1: Time-velocity profile of the HWA measurement on the center line at the pipe outlet of length (a) $L = 0.5$, (b) $L = 3$ and (c) $L = 8$ m at $Re = 4160$

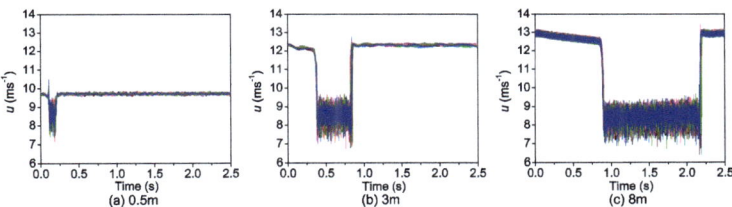

Figure 4.2: Time-velocity profile of the HWA measurement on the center line at the pipe outlet of length (a) $L = 0.5$, (b) $L = 3$ and (c) $L = 8$ m at $Re = 6250$

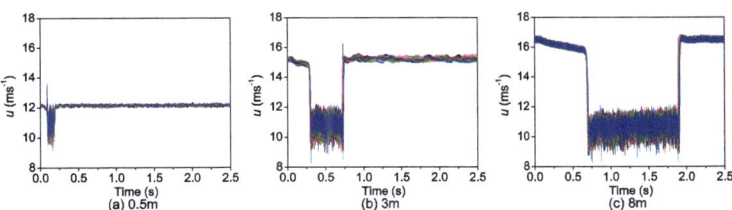

Figure 4.3: Time-velocity profile of the HWA measurement on the center line at the pipe outlet of length (a) $L = 0.5$, (b) $L = 3$ and (c) $L = 8$ m at $Re = 8230$

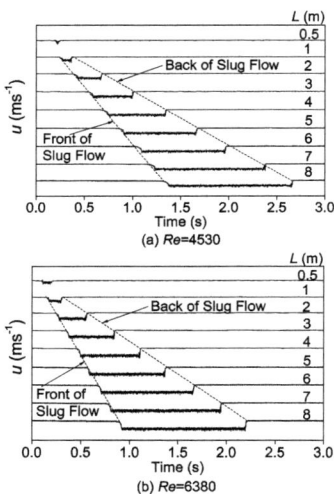

Figure 4.4: Summary of the velocity-time profile of HWA measurement on the center line of pipes of lengths from $L = 0.5$ to 8 m at (a) $Re = 4530$ and (b) $Re = 6380$

65

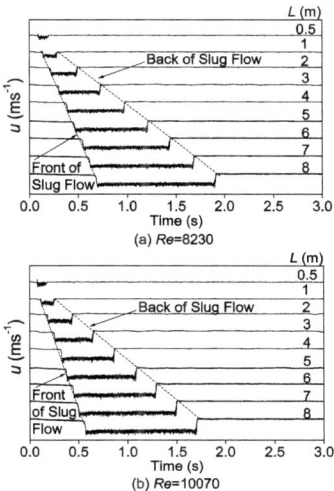

Figure 4.5: Summary of the velocity-time profile of HWA measurement on the center line of pipes of lengths from $L = 0.5$ to 8 m at (a) $Re = 8230$ and (b) $Re = 10070$

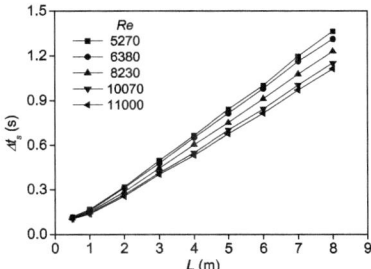

Figure 4.6: Durations of slugs with increasing pipe length for different Reynolds numbers, obtained from the time profile of the HWA measurement

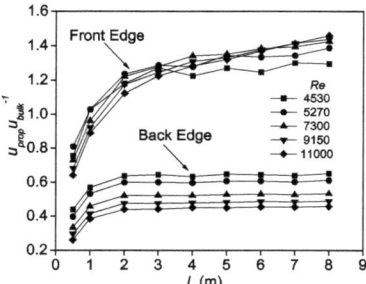

Figure 4.7: Non-dimensionalized slug propagation velocity obtained by division of u_{prop} by the bulk velocity $\bar{u}$ with increasing pipe length from $Re = 4530$ to 11000

realization and the next was chosen as 10 s, because 10 s was sufficiently long for the flow to recover the original laminar flow state after the iris diaphragm was operated. Each of figures 4.1 - 4.3 contains 11 realizations which can no longer be distinguished as an individual realization. The figures show the high reproducibility of triggering by the iris diaphragm system for the entire test rig. During propagation downstream, the slugs' lengths increased, as can be seen from (a) $L = 0.5$ m to (c) $L = 8$ m in figures 4.1 - 4.3. The development of slugs is summarized in figures 4.4 and 4.5, where nine different pipe lengths from $L = 0.5$ to 8 m are shown at Reynolds numbers from $Re = 4530$ to 10070. The figures show clearly that slugs develop in basically the same way at any Reynolds number.

Figure 4.6 provides information on the duration of the deterministically produced slugs depending on how long they propagated downstream. The figure shows that the duration Δt_s of a slug increased linearly with increasing pipe length for almost all pipe lengths except $L \leq 2$ m. For the same length downstream at which slugs were propagated, Δt_s was shorter for high than for low Reynolds number because of the higher bulk velocity.

Since a slug is a portion of a turbulence, the interfaces between laminar to turbulent and turbulent to laminar flow are called the front and back edge of a slug, respectively. Propagation velocities of slugs' front and back edges were also obtained in the following manner. First, the propagation time was measured, i.e. the time between when the iris diaphragm was operated and when the front or back edge of the slug reached the end of the pipe. Second, the pipe length was divided by the propagation time to determine the mean propagation velocity. The propagation velocity is presented in a non-dimensionalized manner by dividing it by the bulk velocity, as shown in figure 4.7. The figure shows that both the front and back edge propagation velocities increase drastically within pipe lengths between $L = 0.5$ and 2 m, which indicates the strong non-linearity of the slug's propagation at the pipe entrance, also caused by the flow development. For $L \geq 2$ m, the back edge propagation velocity does not change with increasing L at any Reynolds number. On the other hand, the front edge continues to increase slightly with increasing L for different Reynolds number and the plots collapse more or less into each other. This shows an apparent difference in nature between front and back edge propagation of slugs.

Usually in natural transitions, slugs occur intermittently, as discussed in section 3.1, originating from some disturbances that exist in the pipe inlet. Therefore, using the present test rig, an attempt was made to simulate such a natural transition situation by applying multiple triggers one after the other as the iris diaphragm system permitted such transitions to be created with arbitrarily chosen time intervals. Consequently, multiple triggering could produce sequences of slugs with certain time intervals between them, as shown in figure 4.8. Figure 4.8(a) and (b) show that the $L = 8$ m pipe was sufficient for the three individual slugs to merge into one large slug with interval times of 0.5 and 1 s, hence one large slug flow left the $L = 8$ m pipe outlet. This is because the front edge of one slug overtook the back edge of another slug while they were propagating downstream. Figure 4.8(c) shows that the pipe length $L = 8$ m was insufficient for the slugs to merge with a triggering interval of 1.5 s. These slugs would have merged, however, if the pipe length had been sufficiently long. This suggests apparently that laminar to turbulent transition phenomena have a strong L/D dependence, which can be expressed by the intermittency γ, defined by equation 3.4. If one takes a certain length of a pipe where transitions occur, γ is higher for the downstream than for the upstream location. The γ dependence on pipe length was first found by Rotta (1956) and the present investigation confirmed it with clear pictures in a systematic way by carefully carried out measurements.

4.2 Puff development in a pipe

It is apparent that the present test rig is also suitable for the introduction of disturbances that result in the formation of puffs. For the investigation of puff development, the iris diaphragm closing height h was set higher than that for slug investigations to ensure that it triggers transitions even at low Reynolds numbers. The Reynolds number of the flow was changed for this set of experiments from $Re \approx 2300$ to 4300, where the transitional structures appear as puffs and also slugs (cf. figure 3.13) and accordingly development of puffs could be studied.

Examples of HWA velocity-time profiles on the center line at the pipe outlet with lengths of $L = 0.5$, 3 and 8 m are shown in figures 4.9 - 4.11 for $Re = 2495$, 2680 and 2865, respectively. Each time record in the figures contains 11 realizations. Unlike the slug results (cf. figures 4.1 - 4.3), realization to realization deviations were observed in figures 4.9 - 4.11, e.g. some transitional structures are completely dissipated, i.e. realizations without showing any velocity drop, as shown in figure 4.9(b) and (c), or some evolve into split puffs, as shown in figure 4.10(b) and (c). In section 4.1, it was clearly shown that the iris diaphragm system had a fairly high reproducibility in triggering transitions, hence those deviations that appear in figures 4.9 - 4.11 are apparently caused by the bifurcation nature of the evolution of puffs in low Reynolds number pipe flows. The issues mentioned above are discussed in detail separately; the dissipation phenomena of the transitional structures in pipe flows are described in chapter 5 and puff splitting is described in section 4.3.

In spite of the existence of deviations in each realization, it was possible to apply the present setup to obtain statistics of puffs by taking large numbers of realizations to find the most frequently occurring form of puffs at various Reynolds numbers. The results of the HWA velocity measurements at Reynolds numbers from $Re = 2310$ to 2865 with different pipe lengths

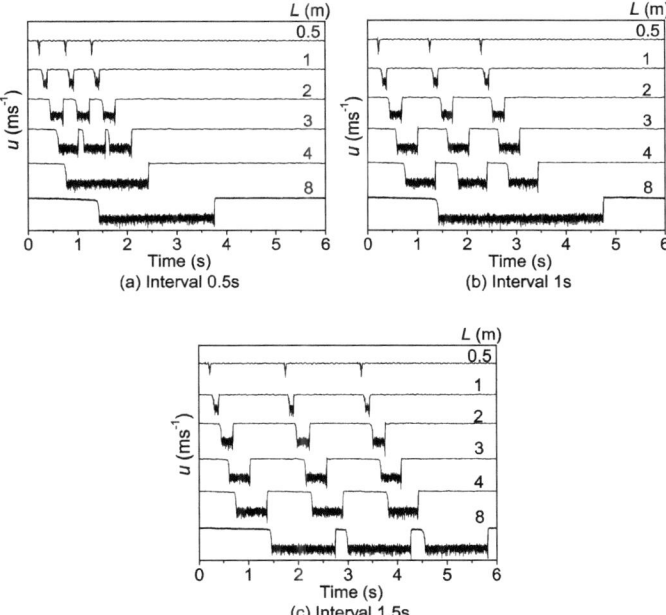

Figure 4.8: HWA measurement on the pipe center line of triple triggering by the iris diaphragm system at $Re = 4200$ with different time intervals: (a) 0.5, (b) 1 and (c) 1.5 s

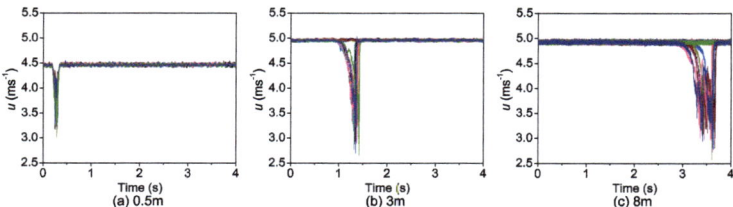

Figure 4.9: Time-velocity profile of the HWA measurement on the center line at the pipe outlet of length (a) $L = 0.5$, (b) $L = 3$ and (c) $L = 8$ m at $Re = 2495$

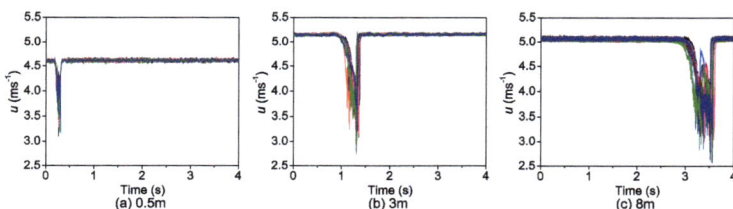

Figure 4.10: Time-velocity profile of the HWA measurement on the center line at the pipe outlet of length (a) $L = 0.5$, (b) $L = 3$ and (c) $L = 8$ m at $Re = 2680$

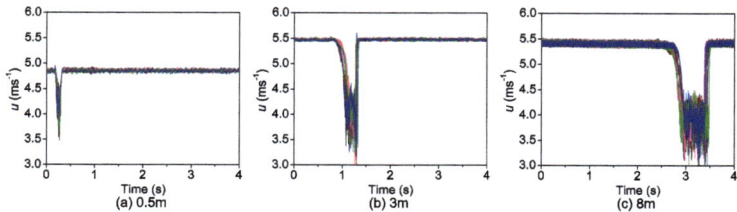

Figure 4.11: Time-velocity profile of the HWA measurement on the center line at the pipe outlet of length (a) $L = 0.5$, (b) $L = 3$ and (c) $L = 8$ m at $Re = 2865$

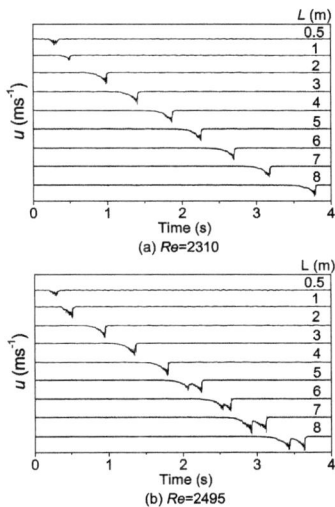

Figure 4.12: Summary of the velocity-time profile of HWA measurement on the center line of pipes of lengths from $L = 0.5$ to 8 m at (a) $Re = 2310$ and (b) $Re = 2495$

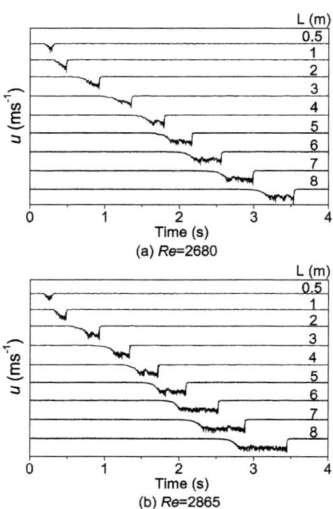

Figure 4.13: Summary of the velocity-time profile of HWA measurement on the center line of pipes of lengths from $L = 0.5$ to 8 m at (a) $Re = 2680$ and (b) $Re = 2865$

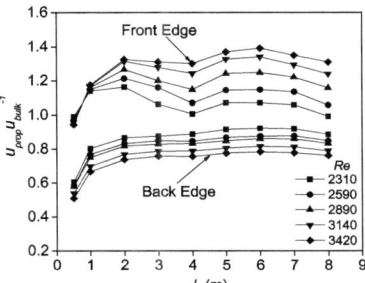

Figure 4.14: Non-dimensionalized puff propagation velocity obtained by division of u_{prop} by the bulk velocity $\bar{u}$ with increasing pipe length from $Re = 2680$ to 2865

are shown in figures 4.12 and 4.13. They show that for the lowest Reynolds number, $Re = 2310$, a single puff was formed in the entrance region of the pipe and it propagated downstream without changing its form. Therefore, such a puff was called first by Wygnanski et al. (1975) an equilibrium puff. When the Reynolds number is increased to $Re = 2495$, an equilibrium puff was available only at about $L = 5$ m and thereafter puff splitting occurred, yielding basically a sequence of two puffs. For even higher Reynolds numbers (see figure 4.13), puff splitting was observed at shorter pipe lengths, permitting the flow between the front and the back edges to develop and yielding slug-like flows further downstream. The front edge of the resulting flow at $Re = 2680$ and 2865 still had the characteristics of puffs, where the back edge of all the resultant flow structures from $Re = 2310$ to 2865 showed a steep change, the same as for the slugs' back edge from the turbulent to laminar state at any Reynolds number, even for equilibrium puffs. Detailed descriptions of the structures of puffs are given in section 4.3.

The propagation velocity of puffs was measured in the same manner for slugs according to the HWA time-velocity profile and presented in a non-dimensionalized way by dividing it by the bulk velocity as shown in figure 4.14. The front and back edge propagation velocities of the puffs did not change much with increase in pipe length for $L \geq 2$ m, whereas the front edge propagation velocity of slugs continued to increase, and the back edge propagation velocity remained more or less constant as in the case of slugs. Both the front and back edges of puffs show propagation velocities very close to the bulk velocity, especially for the lower Reynolds number range, i.e. the (equilibrium) puff is localized and propagated downstream.

The propagation velocities of puffs and slugs are summarized in figure 4.15 for all the investigated Reynolds number range. Results from the literature (Lindgren (1969), Wygnanski *et al.* (1975) and Durst & Ünsal (2006)) are also plotted for comparison. Previous measurements of the propagation velocity was carried out with one long pipe test rig by visualization (Lindgren (1969)) and pressure measurements through pressure tabs bored directly in the pipe wall (Wygnanski *et al.* (1975), Durst & Ünsal (2006)). In contrast, the present investigations were carried out by velocity measurements at the pipe outlet of different lengths. The results of the present investigations and the previous studies merge with each other, which indicates that the propagation of puffs and slugs does not depend on the downstream conditions, e.g. pipe length. The front edge propagation velocity increases continuously and the back edge propagation velocity decreases continuously with increase in Reynolds number. Whether the propagation velocities would converge to a finite value with increase in Reynolds number should be studied by further investigations at higher Reynolds number.

4.3 Puff to slug transformation

As clearly presented in section 4.2, a transformation from puffs to slugs is initiated by puff splittings, with increase in pipe length or/and in Reynolds number. To gain a deeper insight into the transformation phenomena, a realization-averaged velocity-time profile was taken for different structure forms found at different Reynolds number, as presented in figure 4.16, which shows (a) an equilibrium puff, (b) a split puff, (c) a not fully developed slug and (d) a fully developed slug. Each realization-averaged velocity-time profile is a typical transitional flow structure at each Reynolds number, showing its typical characteristics as follows. (a) An equilibrium puff does not have a

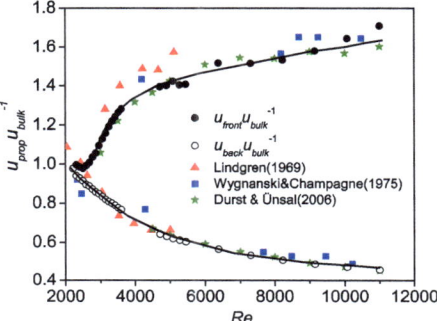

Figure 4.15: Non-dimensionalized propagation velocities of the front and back edges of transitional flow structures compared with previous results in the literature

clear interface from laminar to turbulent in the front edge but a clear interface from turbulent to laminar in the back edge. (b) A split puff has similar front and back edges to a typical equilibrium puff, but in the middle it splits into two parts which are a sequence of two equilibrium puffs combined. (c) A not fully developed slug has a front edge still similar to that of equilibrium puffs but the back edge is same as that of an equilibrium puff. (d) A fully developed slug has a clear interface at the front and back edges.

Taking the realization average of each kind of structures allowed the calculation of the turbulent intensity Tu for each, as shown in figure 4.17. Figure 4.17(a) - (d) show Tu values corresponding to the cases illustrated in figure 4.16(a) - (d). Figure 4.17(a) reveals that Tu varies from 2% to 15%, which is the case for the equilibrium puff ($Re = 2310$). The split puffs show a similar trend of variation in Tu as two equilibrium puffs combined, shown in figure 4.17(b). Figure 4.17(c) and (d) show that the structures have similar aspects to each other, keeping a maximum intensity value close to 15%, and the Tu values in the middle of the structures are 5% and 4.3%, respectively. For $Re \geq 4345$, the Tu values of the slugs' internal structures are constant at $Tu \approx 4.3\%$, which is a typical value for fully developed turbulent flows.

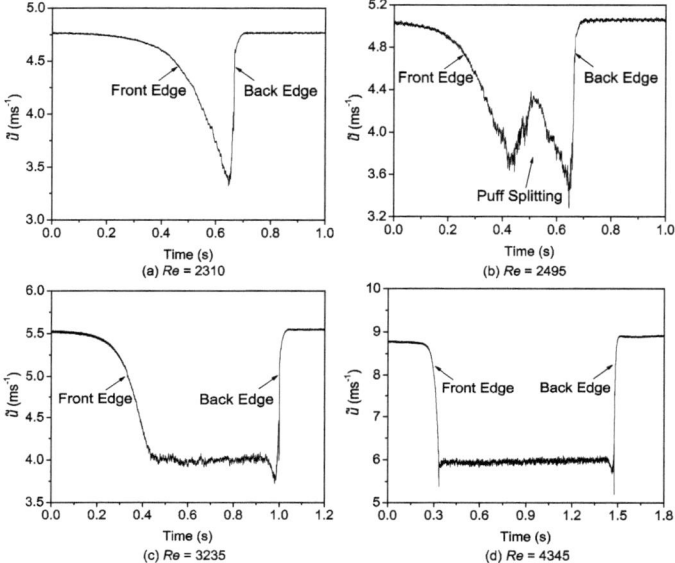

Figure 4.16: Realization-averaged HWA velocity-time profile measured on a center line at Reynolds numbers (a) $Re = 2310$, (b) $Re = 2495$, (c) $Re = 3235$ and (d) $Re = 4345$ with pipe length $L = 8$ m

Hence it was possible to define the differences in those structures, being a fully developed slug or a not fully developed slug, by measuring the Tu of the internal structures. The slug (internal) structure was investigated more in detail with measurements of turbulent Reynolds stress anisotropy components, presented in chapter 7.

To gain an insight into the structures of puffs, the velocity-time profile was obtained for different positions in a pipe cross-section at the $L = 8$ m pipe outlet. The results shown in figure 4.18 are ensemble-averaged values for 60 realizations at different radial locations r/R, where r is the radial distance from the center line of the pipe and R is the pipe radius. The puff structures

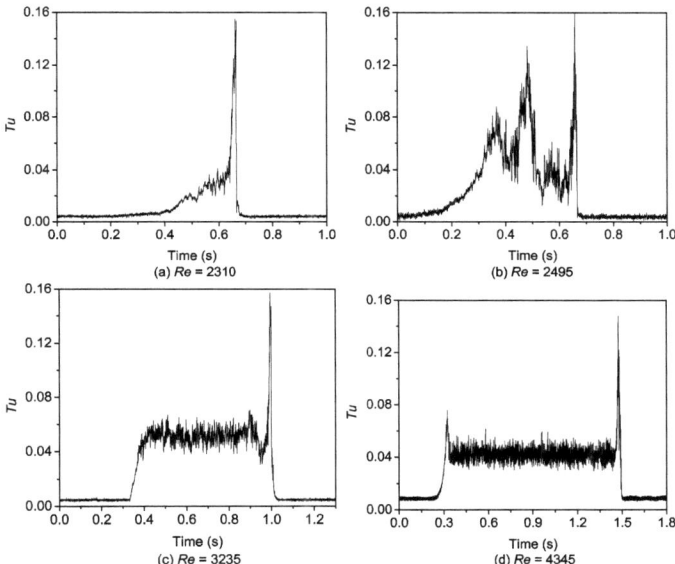

Figure 4.17: Turbulent intensity measured on a center line at Reynolds numbers (a) $Re = 2310$, (b) $Re = 2495$, (c) $Re = 3235$ and (d) $Re = 4345$ with pipe length $L = 8$ m

presented in figure 4.18, corresponding to $Re = 2450$, do not have a block profile inside them whereas slugs do, as shown by e.g. Durst & Ünsal (2006). At the location $r/R = 0.93$, the axial velocity does not show a large change even when the velocity at the center line ($r/R = 0$) reaches the maximum deviation from the laminar state. This complex structure of puffs was discussed by van Doorne & Westerweel (2007), who visualized puffs by the SPIV system. They mentioned that this is an indication of the quasi-periodic regeneration of hairpin vortices. The present results show good agreement with their observations; whereas they took only one puff structure, the present results were taken as realization-ensemble-averaged signals. The results shown in figure 4.18 are replotted in a different manner to present typical cross-sectional velocity profiles of puff structures at $Re = 2450$, as shown in figure 4.19 where the velocity profile is shown between $t = 4$ and 4.4 s in a symmetrical manner for clear presentation. The figure shows the existence of a vortex-like structure near the wall region. Dou (2006) and Wedin & Kerswell (2004), for example, predicted the velocity profile during transitions in pipe flows and reported that $r/R \approx 0.6$ is the point of breakdown initiation. Figures 4.18 and 4.19 show that the velocity-time profile at $r/R \approx 0.6$ gives signals with the highest amplitude in fluctuations, which may be caused by the initiation of breakdown as reported in the literature.

Wygnanski & Champagne (1973) showed two clearly separated regions for the appearance of slugs and puffs in transitional pipe flows. They observed puffs for very high levels of disturbances and at low Reynolds numbers as in the present investigations, and mentioned the existence of puff splitting. This clear separation between slugs and puffs was not found by Darbyshire & Mullin (1995), who detected both puffs and slugs in some flow regimes. They also found that this mixed occurrence was not just dependent on the magnitude of the disturbance but also on the type of flow disturbances that they introduced. In the present study, the flow disturbance was introduced at the pipe inlet by a short duration of insertion of a flow blockage with the blockage area being kept constant. As the Reynolds number was increased, first equilibrium puffs, then split puffs and multiple split puffs were observed, that finally yielded flow structures that have an overall appearance similar to that of slugs.

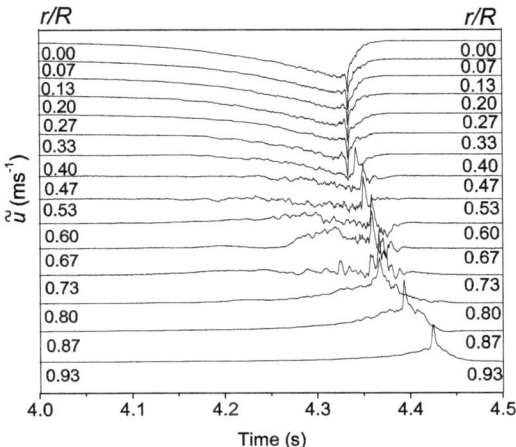

Figure 4.18: HWA velocity-time profile at different radial positions r/R at times from 4 to 4.5 s after the iris diaphragm is operated at $Re = 2450$

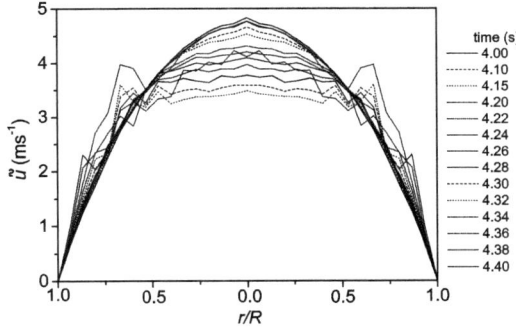

Figure 4.19: Axial velocity as a function of different radial positions r/R for times from 4 to 4.4 s after the iris diaphragm is operated at $Re = 2450$

It was observed from the above-mentioned results that it is necessary for puff splitting to occur during the transformation from puffs to slugs. In a certain Reynolds number range, one can see various transitional flow structures with the same triggering at the same Reynolds number. Figure 4.20 shows example signals of transitional flow structures that appeared at $Re = 2570$ with triggering by the iris diaphragm system with constant blockage area and lapse time. Figure 4.20 shows (a) an equilibrium puff, (b) a split puff, (c) a triple split puff and (d) a multiple split puff. It is interesting that for all structures the back edge reaches the pipe outlet at around a same time of 3.75 s but the front edges reaches the pipe outlet at different times. The more the split occurs, the faster the front edge propagates downstream, i.e. puff splitting made its propagation suddenly fast in the downstream direction but not at all in the upstream direction. The propagation velocity of the front edge of a slug is typically much faster than the bulk velocity, as shown in section 4.1, hence figure 4.20 confirms that the puff splitting is the initiation of the transformation from puffs to slugs. Figure 4.20 also suggests a complex bifurcation nature during the evolution of puffs in the pipe because although the triggering setting and the flow conditions are kept the same, a triggered flow structure may remain as an equilibrium puff or break down into a single or multiple split puff. The developments of an equilibrium puff and a split puff were shown in figure 4.12 at $Re = 2310$ and 2495, respectively. They propagate downstream as a localized structure but, at sufficiently high Reynolds number, they do not maintain their structure and break up into multiple split puffs while propagating downstream. The pipe location where puff splitting occurs depends on the Reynolds number, hence it is of interest to investigate the dependence on the pipe lengths where equilibrium puffs turn into split puffs.

Investigations on the transformation from puffs to slugs were carried out further by obtaining the number of puff split occurrences with increase in Reynolds number, which is presented in the form of a probability curve as shown in figure 4.21. The probability is taken such that the number of realizations of split or multiple split puffs (e.g. shown in figure 4.20(b) - (d)) is divided by the total number of realizations. It thus represents the appearance frequency of split puffs at the end of the $L = 8.5$ m pipe. The curve is similar

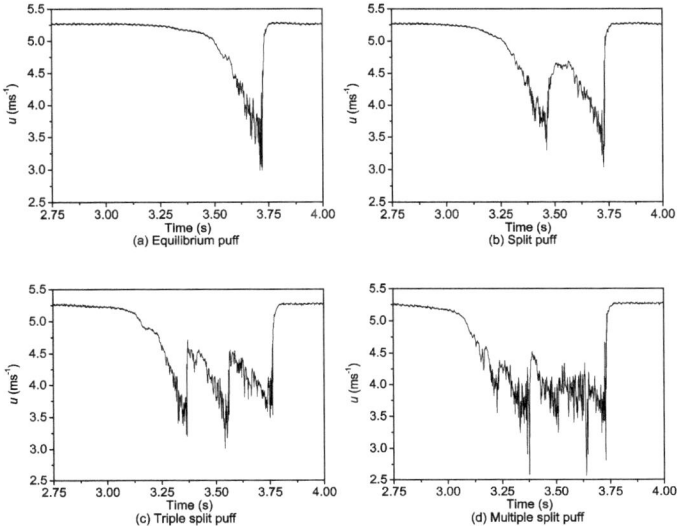

Figure 4.20: Examples of varieties of transitional flow structures at $Re = 2570$: (a) equilibrium puff, (b) split puff, (c) triple split puff and (d) multiple split puff; HWA velocity-time profiles measured on a center line with s pipe length $L = 8.5$ m

to the intermittency curve of puffs (cf. figure 3.9) and indicates that the puff splitting started to appear at $Re = 2325$ and all the puffs were split at $Re \geq 2650$.

4.4 Summary of development of puffs and slugs

In the present chapter, the development of puffs and slugs in a pipe was first investigated at various Reynolds number, using an iris diaphragm system to trigger the flow in well-controlled manner. The closing area and the lapse time of the iris diaphragm were varied occasionally so that transitional flow structures could be deterministically studied at targeted Reynolds numbers

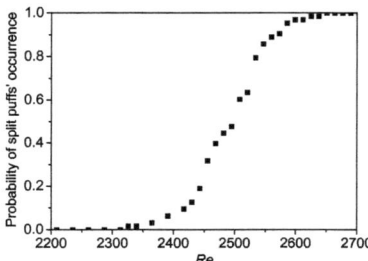

Figure 4.21: Probability curve of the occurrence of puff splitting detected by the HWA at the end of an $L = 8.5$ m pipe

for $Re < Re_{crit}$. First, it was shown how slugs developed in a pipe. It was found that they developed in more or less the same manner at any Reynolds number. The propagation velocities of the front and back edges of slugs were measured and it was found that the propagation velocity of the front edge increased continuously with increase in pipe length and also in Reynolds number. On the other hand, the back edge propagation velocity did not continue to increase with increase in pipe length but continued to decrease with increase in Reynolds number. After that, the natural transitions were simulated by multiple triggering of the flows, with results presented as the L/D dependence on γ of transitional pipe flows.

Second, the development of puffs in pipe was comprehensively demonstrated. The evolution of puffs while they propagated downstream varied, hence the phenomena should be treated as a statistical manner by taking the average of a large number of realizations. It was shown that puffs' structures were strongly dependent on Reynolds number as dissipations and splits of puffs were observed, depending on the Reynolds number. The measurement of the propagation velocity of puffs showed apparently that an equilibrium puff is, and a split puff can also be, a localized structure. Different transitional flow structures at low Reynolds number were described precisely by realization-averaged velocity signals and Tu values.

Finally, the transformation from puffs to slugs was apparently shown.

It was demonstrated that a puff splitting was the initiation of the puff to slug transformation. Puff splitting occurred in a certain range of Reynolds number at certain position in a pipe. Various kinds of puffs could be observed under the same flow conditions and triggering setting, which gave additional proof that puff splitting was the initiation of the transformation from puffs to slugs. The systematic relation between the number of puff split occurrences and increase in Reynolds number was shown by the occurrence probability curve, which suggests further investigations to study, e.g., the location of the occurrence of puff splitting, depending on Reynolds number.

Chapter 5

Lifetime of transitional flow structures

As introduced in section 4.2 in figure 4.9, some transitional flow structures completely dissipated in the pipe while they propagated downstream, although the flow was indeed triggered. The present chapter describes studies focused on the dissipation phenomena of transitional flow structures in pipe flows, which can be observed only at low Reynolds number, in parallel with the available knowledge explained above. For this purpose, a simple pressure-drop measurement method is presented for the direct measurement of the 'full-lifetime' of transitional structures in low Reynolds number pipe flow, which is the time required for the transitional structures to dissipate fully just after disturbing the flow with constant amplitude and duration, as shown in figure 4.9(b) and (c), where some triggered flows were dissipated in a pipe completely while propagating downstream. It should be noted that the transitional flow structures at low Reynolds numbers involve those structures evolving to a puff, developed puffs, split puffs and slugs. It is already clear from the investigations described in section 4.2 that the transitional phenomena in pipes occurring at low Reynolds numbers are stochastic. Therefore, sufficient numbers of repetitions of the measurements were made to determine the most frequently occurring phenomena and also to obtain reliable statistics. Additionally, one could even trace the development and also the dissipation of transitional flow structures from the pressure-drop signals obtained, hence it is precisely described in this chapter how transitional flow structures in low Reynolds number pipe flows evolve in a pipe. In addition to direct full-lifetime measurements, hot-wire measurements were also conducted at the pipe outlet to monitor the velocity form of transitional structures and complement the direct full-lifetime measurements with probability of occurrence analysis and determination of the lifetime. These are explained in section 5.1.

Section 5.2 gives a detailed account of the findings relevant to full-lifetime measurements for $Re < Re_{crit} = 1880$, which is defined as the critical Reynolds number at which the disturbed structures start to develop into sustained puffs. The measurements were conducted in pipes having lengths of 200, 300 and 566 pipe diameters after the location of induced disturbances. The results of measurements conducted for the present investigation, which show bifurcation phenomena observed in the full-lifetimes, are also explained. In section 5.3, the results are summarized and discussed, suggesting possible evolution routes for transitional structures in pipe flows at low Reynolds number.

5.1 Method of direct measurement of lifetime

For the investigation of the lifetime of transitional flow structures, the experimental test rig used in the previous investigations was partially modified with an alternative mass flow rate controller unit, MFCU2, to achieve a higher controlling resolution of Reynolds numbers, as mentioned in chapter 2. The modified experimental test rig is shown schematically in figure 5.1. For the sake of completion, a brief description of the test rig employed follows here. A mass flow rate controller (Bronkhorst, F-202AC-FB-44V) is employed, which has a deviation of less than 0.4% from the required mass flow rate value. The controller is connected to a 5 bar pressurized air supply and it allows the Reynolds number to be adjusted in steps of 5 in the present setup. A critical nozzle is connected to the mass flow controller in order always to have choked flow and, consequently, it was able to maintain a constant mass flow rate independent of the pressure drop caused by the occurrence of transitions in the downstream region of the pipe. A flow condi-

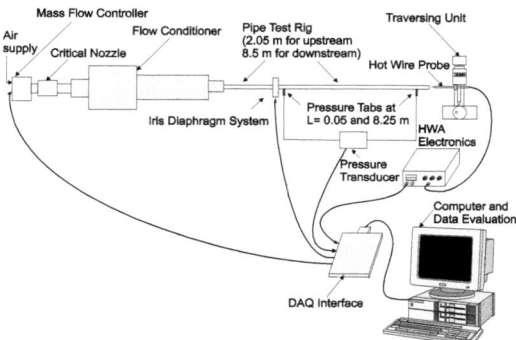

Figure 5.1: Schematic diagram of the experimental test rig with an iris diaphragm system for the direct lifetime measurement of transitional flow structures

tioner is installed prior to the 15 mm diameter pipe to suppress irregularities in the flow and acoustic disturbances. The critical Reynolds number of the naturally occurring transition of the present system was around $Re = 13000$.

To trigger transition, an iris diaphragm system was employed (figure 5.1). The system was able to control the closing area and also the lapse time of the iris diaphragm. It was set at constant values for the closing area of 30% and for the closing time of 40 ms for most of the experiments. The iris diaphragm was placed at $L = 2.05$ m ($137D$) downstream of the flow conditioner, so that a flow could develop into Hagen-Poiseuille flow up to $Re \approx 2440$ (cf. equation 3.1), before it was disturbed by the iris diaphragm. Pipes with various lengths are connected after the iris diaphragm system to obtain the probability of occurence curves. The maximum length of the pipe was 8.5 m ($566D$) in the present experiments.

An HWA was employed to measure the center line velocity at the pipe outlet and a piezo-resistive pressure sensor (RCI, PD 2915 SM1) was employed to measure the pressure difference Δp between $L = 0.05$ and 8.25 m ($3D$ and $550D$) after the triggering location. Figure 5.2 shows an ex-

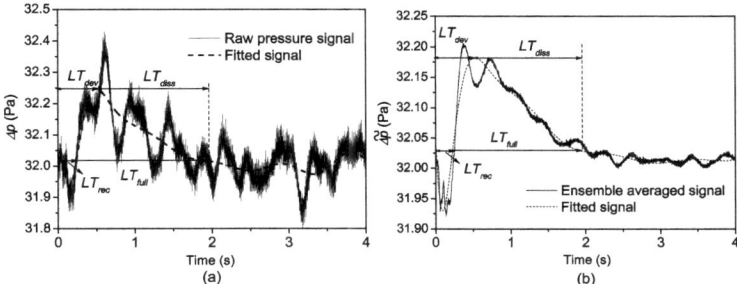

Figure 5.2: Example of direct lifetime measurement by pressure difference and the definition of the various characteristic times: (a) raw pressure signals and (b) an ensemble-averaged signal (at $Re = 1850$)

ample of the time variation in the pressure-drop signal, where the x-axis is the time after the iris diaphragm operated and the y-axis is the pressure difference. Figure 5.2(b) depicts the ensemble-averaged pressure signal $\widetilde{\Delta p}$ over 200 realizations. The increase and decrease in pressure difference indicate the development and the dissipation of transitional structures at the set Reynolds number. Thus, the elapsed time between the operation of iris diaphragm ($t = 0$) and the pressure difference returns back to the original value is defined as the full-lifetime, LT_{full}. Moreover, recovery time, LT_{rec}, development time, LT_{dev}, and dissipation time, LT_{diss}, are also defined in figure 5.2. LT_{rec} is the time for the pressure to recover from its initial drop due to the opening of the iris diaphragm. The time the transitional structures took to reach the maximum value is termed LT_{dev}. LT_{diss} is defined as the time for the decay of the pressure signal. Hence the following relation holds between the definitions of characteristic times:

$$LT_{full} \approx LT_{dev} + LT_{diss}. \tag{5.1}$$

The LT_{rec} and LT_{dev} are characteristic times which are supposedly dependent on the flow facility and the type of disturbance.

One can observe low-frequency oscillations in the pressure signal in figure 5.2. Such oscillations were also seen in the hot wire anemometry signal.

Hence the oscillations originated from a low-amplitude fluctuation of the bulk flow. This is due to the system characteristics of the mass flow controller, but since the measurement range of pressure difference is already less than 0.5 Pa, the oscillation looks large but the resulting deviation in Reynolds number was within ±0.5%. Therefore, it is expected that the influence of these fluctuations on the general behavior of transitional structures is negligible.

The procedure for direct measurement of LT_{full} of transitional structures was as follows:

1. Set the flow rate to the mass flow controller according to a chosen Reynolds number

2. Operate the iris diaphragm to trigger the flow

3. Measure the center line velocity at the outlet of the pipe and also the pressure difference between two points, L=0.05 m (3 D) and 8.25 m (550 D).

4. Pick up the pressure signals of realizations for which the HWA did not show the velocity drop (indication of a puff at the pipe outlet).

5. Take an ensemble average of pressure signals, picked up in step 4, and evaluate the full-lifetime from the mean signal.

6. As an alternative to step 5, for each pressure signal picked up in step 4, make curve fits and extract values of all characteristic times from the curve fits and analyze their mean values.

By the above-described method, one can only measure the full-lifetime of transitional structures which were completely dissipated in the pipe. If they did not dissipate, they would appear as puffs detected by the HWA at the pipe outlet with a lifetime beyond the measurable range. At least 60 realizations were taken for each Reynolds number for $Re < 1830$, but over 200 realizations were sampled for $Re \geq 1830$ to obtain good ensemble averaging. The time between each triggering was chosen as 10 s, for the same reason as in the previous investigations that 10 s was sufficiently long for the flow to recover the original laminar flow state after the iris diaphragm operation.

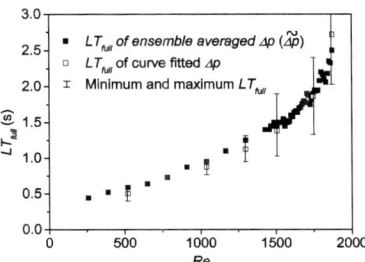

Figure 5.3: Comparison of LT_{full} evaluated with two different methods of analyzing the transient pressure signal for $L = 566D$

LT_{full} values determined from the ensemble-averaged signal (step 5) and from curve fits (step 6) match well, as shown in figure 5.3. The analysis made by curve fits allowed the scatter of LT_{full} to be observed, which increases substantially for $Re > 1550$. In the subsequent analysis, LT_{full} evaluated from the ensemble-averaged signal is employed. The lifetimes extracted from curve fits in figure 5.3 had a distribution such that 80% of all realizations showed their lifetime within ±10% of the average value.

In figure 5.4(a), the changes in all kinds of lifetime are shown for pipe length $L = 566D$. As can be seen, LT_{rec} and LT_{dev} remain constant over the whole Reynolds number range with values of ~ 0.22 s and ~ 0.4 s, respectively, whereas LT_{full} and LT_{diss} increase with increase in Reynolds number. LT_{full} measurements conducted within pipes having different lengths do not show any systematic deviation from each other as can be seen in figure 5.4(b), i.e. the pipe length has no influence on the lifetime measurements, when a transitional structure does not reach to the pipe outlet. In other words, the lifetime measurement through Δp was not affected by the downstream conditions, such as pipe length. The independence of the downstream condition on the evolution of transitional flow structures was briefly mentioned in section 4.2 (figure 4.15).

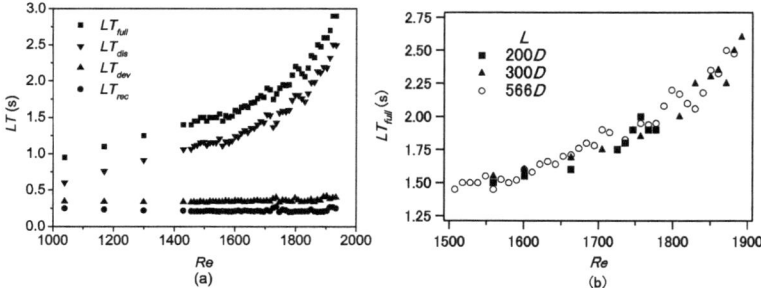

Figure 5.4: (a) Variation of different types of LT with Re for $L = 566D$; (b) LT_{full} measured within pipes having three different lengths.

5.2 Analysis of lifetime and probability measurements

Figure 5.5 shows a probability curve of the occurrence of puffs detected by the HWA at the pipe outlet having lengths $L = 200D, 300D$ and $566D$. The probability was obtained by dividing the number of realizations for which puffs were detected by the HWA by the total number of realizations. For $L = 566D$, the probability curve starts to increase from $Re = 1890$ and reaches 1 at around $Re = 2080$, i.e. no realization was detected as a puff for $Re < 1890$ and all realizations were detected as puffs for $Re \geq 2080$ by the HWA at the pipe outlet. Thus, according to figure 5.5, all the triggered structures were completely dissipated in the pipe with $L = 566D$ at $Re < 1890$. Although there were some structures which were dissipated in the pipe at $Re \geq 1890$, some of them were sustained longer than the pipe length, hence they were detected by the HWA as puffs.

LT_{full} in figures 5.4(a) and 5.6 shows a gradual increase with increase in Reynolds number, then there is a 'plateau' of the curve in the range $1450 \leq Re \leq 1550$ and, subsequently, it starts to increase rapidly. Note that LT_{full} is plotted in figure 5.6 only for $Re < 1890$, because above this Reynolds number puffs start to live longer than the pipe length. Moreover,

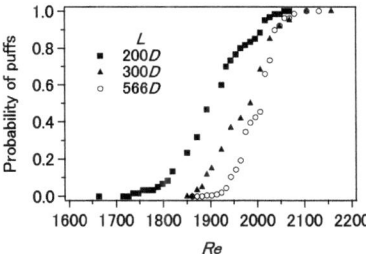

Figure 5.5: Probability curve for the occurrence of puffs detected by the HWA at the pipe outlet of three different lengths

pressure signals suddenly started to show fluctuations with higher amplitudes and realizations show considerable differences from one realization to the other at $Re = 1560$ compared with the Δp signals for $Re \leq 1550$ (figure 5.7). The results in figure 5.6 showed apparently that there was a discontinuity in the increase in lifetime with increase in Reynolds number at around $Re \approx 1550$, hence one can say that this is the Reynolds number which is the boundary between two distinct transitional flow structures, namely 'just triggered' structures and 'developed' structures. In other words, at around $Re \approx 1550$, the evolution from 'just triggered' structures into 'developed' structures could be recognized. In the realizations of $Re = 1560$ in figure 5.7(b), deviations are observed at time 0.5 s, which is immediately after the iris diaphragm was operated. This represents the beginning time and location of the evolution of a transitional flow structure. This value is fairly close to the results of visualization reported by Peixinho & Mullin (2007), who found that 'just triggered' structures started to show a disorder only after the front edge moved $4D$ (in time, 0.6 s).

Figure 5.8(a) shows Δp signals of seven realizations at $Re = 1880$. As can be seen, most of the realizations reach the maximum value of Δp, and decrease immediately. Only one signal in figure 5.8(a) shows a different tendency, i.e. first it increases similar to those in the other realizations but then it continues to increase further and subsequently decreases. As a con-

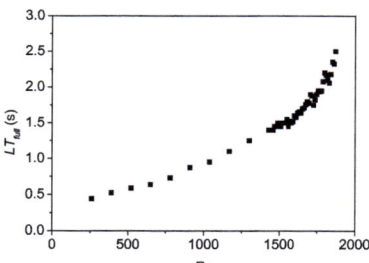

Figure 5.6: Change in LT_{full} of transitional flow structures with Re in the pipe with $L = 566D$.

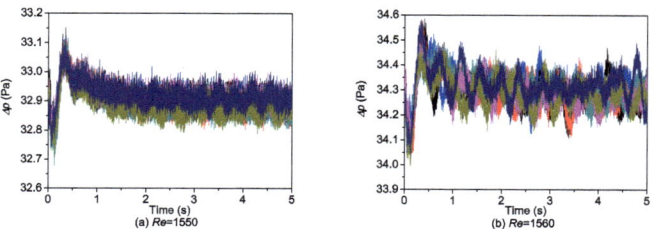

Figure 5.7: Δp of seven distinct realizations in the pipe having $L = 566D$ at (a) $Re = 1550$ and (b) $Re = 1560$

sequence, these kinds of transitional flow structures have different lifetimes. It should be noted that the deviations from one realization to the other was not originated by the triggering, because the iris diaphragm system already showed high reproducibility in previous investigations. Therefore, the two different tendencies of the structures shown in figure 5.8 can be accepted to be the bifurcation in the evolution of the disturbances. This kind of large deviation from one realization to the other appeared suddenly at $Re = 1880$, which is defined in the present investigation the *critical Reynolds number* (Re_{crit}). Willis & Kerswell (2007) found a 'crisis Reynolds number' as $Re = 1870$ by direct numerical simulation, where suddenly the transitional flow structures start to have a long lifetime. Their 'crisis Reynolds number' is fairly close to that found in this study.

In the following discussion, the transitional flow structure that decays immediately after it reaches a peak Δp, as for most of the signals shown in figure 5.8(a), is called a 'directly dissipated' structure; the single, long-lasting signal Δp shown in the same figure is called a 'decayed sustained' structure. The term 'decayed' in the name was necessary because non-decayed 'sustained' structures were also observed, which propagated further downstream, longer than the pipe length, and were therefore detected as puffs by the HWA. Hence there are two 'sustained' structures, and the lifetimes of only decayed ones could be measured, and accordingly are called 'decayed sustained' structures. Compared with 'directly dissipated' structures, 'sustained' structures have typical characteristics: they need a longer time to develop and they reach a higher maximum value of Δp. For $Re \geq 1890$, the HWA starts to detect puffs where Δp showed exactly the same characteristics as a 'decayed sustained' structure, taking a longer time to develop than a 'directly dissipated' structure, and decrease because transitional flow structures reach the pipe outlet. As the Δp signals 'directly dissipated' and 'decayed sustained' are considerably different from each other, one can clearly distinguish two ensemble-averaged Δp signals for 'sustained' and 'directly dissipated' structures. Figure 5.8(b) shows typical ensemble-averaged Δp signals for 'directly dissipated' structures and also 'decayed sustained' structures at $Re = 1945$.

With increase in Reynolds number, the number of 'sustained' structures increased. Nevertheless, 'directly dissipated' structures also existed at the

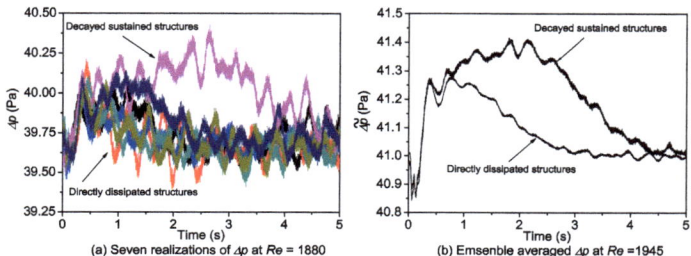

Figure 5.8: (a) Δp-time profile of seven realizations at $Re = 1880$ and (b) ensemble-averaged Δp at $Re = 1945$

same Reynolds number. Therefore, it is necessary to discuss the occurrence ratio of each transitional flow structure that appeared at a Reynolds number for $Re \geq Re_{crit}$, namely 'directly dissipated', 'decayed sustained' and 'sustained' structures. There are certain ratios in their occurrence depending on the Reynolds number. Figure 5.9 shows the occurrence probability curves for each structure, which were calculated as the number of realizations of each structure divided by the total number of realizations. Figure 5.9 shows that the occurrence probability of 'directly dissipated' structures decreases with increase in Reynolds number and for $Re \geq 2080$ no 'directly dissipated' structures exist. The occurrence probability of 'decayed sustained' structures increases with increase in Reynolds number together with 'sustained' structures (puffs). The results reveals that, for example, at $1880 \leq Re \leq 1970$, more than 50% of tansitional structures dissipate directly.

'Directly dissipated' structures and also 'decayed sustained' structures exist only up to $Re = 2080$, because the occurrence probability of puffs (sustained structures) reaches 1 at $Re = 2080$, as shown in figure 5.9. The existence of 'directly dissipated' structures up to $Re = 2080$ gives the reason why the puff occurrence probability curves for the three pipes with different lengths in figure 5.5 start to increase at different Reynolds numbers but reach 1 at more or less the same Reynolds number. The three probability curves in figure 5.5 show similarity only for the lower range of probability. The

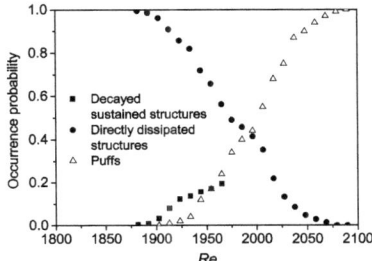

Figure 5.9: Occurrence ratio of three bifurcated transitional flow structures for $Re > Re_{crit}$

probability curve for the occurrence of puffs for an $L = 200D$ pipe covers a wider range of Reynolds number, about 350, than that for an $L = 566D$ pipe, about 200. This is because of the existence of 'directly dissipated' structures up to $Re \leq 2080$.

It is obvious that the lifetimes of transitional flow structures for $Re \geq 1890$ show a bifurcated nature during the evolution in the pipe, one being dissipated and the other sustained. Hence the lifetimes extracted from the ensemble-averaged signals, which were evaluated separately for the 'directly dissipated' and 'decayed sustained' structures similar to those in figure 5.8(b), show two possible lifetimes at a Reynolds number $Re \geq Re_{crit}$ (figure 5.10). Moreover, the lifetime of 'directly dissipated' structures does not increase with increase in Reynolds number but remains constant at around 2.6 s. The lifetime of 'decayed sustained' structures increases with increase in Reynolds number, a tendency that is similar to that for for $Re < Re_{crit}$. It was possible to measure the lifetime of 'decayed sustained' structures only up to $Re \leq 1964$ since all 'sustained' structures reach the pipe outlet and appear as puffs. However, if the pipe length were sufficiently long, the increase of the lifetime of 'decayed sustained' structures with Re could be measured.

It has been reported by, e.g., Peixinho & Mullin (2006) and Faisst & Eckhardt (2004) that the lifetime of transitional flow structures in pipe flows should be proportional to $(Re_{crit} - Re)^{-1}$, hence the relations between life-

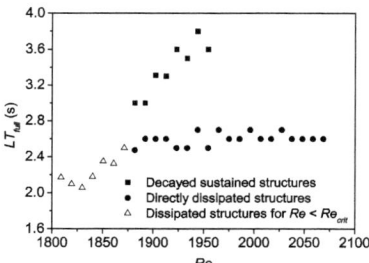

Figure 5.10: Change in LT_{full} with increase in Reynolds number for $Re > Re_{crit}$

times LT_{full}^{-1} and LT_{diss}^{-1} and the Reynolds number are plotted in figure 5.11. Figure 5.11(a) shows that LT^{-1} decreases exponentially for $Re \lesssim 1500$, then it remains constant up to $Re \approx 1550$ and subsequently it decreases linearly with increase in Reynolds number for $1500 \gtrsim Re \leq 1880$. It can be clearly seen from the linear trend lines in figure 5.11(b) that both LT_{full}^{-1} and LT_{diss}^{-1} have a tendency to approach zero for $Re > 1500$, which suggests that the lifetime of transitional flow structures would be infinite at a finite Reynolds number. The intercept point of the trend lines on the Reynolds number axis is around 2300. As the lifetime diverges rapidly for $Re > Re_{crit}$, owing to the necessary pipe lengths it is almost impossible to conduct direct full-lifetime measurements and simulations beyond Re_{crit} and confirm this trend. Nevertheless, it is also difficult to consider that there is an infinite lifetime of transitional flow structures at low Reynolds numbers as figure 5.11 suggests. Interestingly, the breakdown of a puff occurs at $2300 \leq Re \leq 2700$, as shown in figure 4.21, where the divergence of LT of single puffs is expected (vanishing LT^{-1}). This observation leads to the conclusion that transitional structures favor breakdown instead of having extremely long or infinite LT.

The effects of disturbance type and amplitude on the probability of occurrence of puffs and derived lifetimes from the probability curves were shown by Mullin & Peixinho (2006), who concluded that the lifetime depends on the type of disturbance. In order to establish the effect of disturbance ampli-

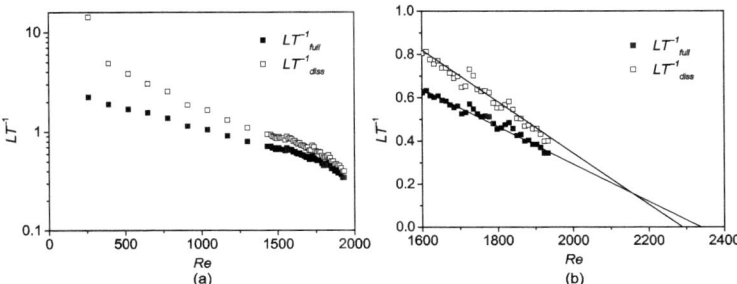

Figure 5.11: (a) Change in LT_{full}^{-1} and LT_{diss}^{-1} with the increase in Reynolds number; (b) trend of LT_{full}^{-1} and LT_{diss}^{-1} based on the measured data in the Reynolds number range $1550 < Re < Re_{crit}$

tude on the directly measured LT_{full}, one set of experiments with a different disturbance amplitude was carried out in the pipe with $L = 566D$. The new disturbance amplitude of the iris diaphragm system was set as a closing area of 40% and a lapse time of 50 ms, which is larger than the previous one (30% and 40 ms, respectively). The comparison of the puff occurrence probability curves for two different applitudes in figure 5.12 reveals the effect of disturbance amplitude in the probability statistics: the probability curve for the 40% amplitude disturbance starts to increase at $Re \approx 1920$ and reaches 1 at $Re \approx 2080$ and csubsequently shifts toward a slightly higher Reynolds number region than that with 30% amplitude disturbance. Hence it can be argued that the flow against the lower disturbance amplitude is more stable than that against the higher disturbance amplitude, which indicates the complexity of the relation between the disturbance amplitude and the triggering, development and decay of transitional flow structures. Similar effects related to the nonlinearity of disturbance amplitudes were also observed during previous investigations with ring obstacles described in sections 3.2 and 3.4 and by Mullin (2008). The LT_{full} of transitional flow structures triggered by the higher amplitude collapse to those measured with lower disturbance amplitude, as shown in figure 5.12(b). In contrast to the dependence on lifetime

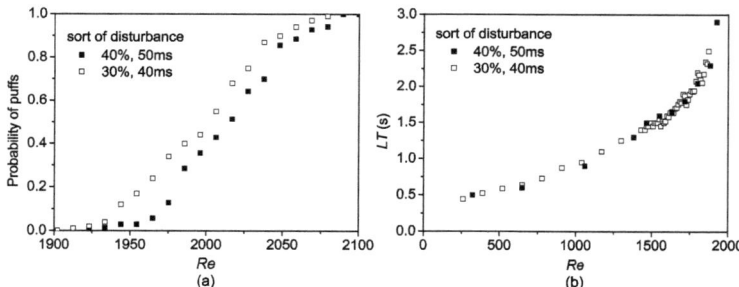

Figure 5.12: (a) Probability of occurrence of puffs detected by the HWA with two disturbances of different amplitudes; (b) LT_{full} of transitional flow structures triggered by two disturbances of different amplitudes

evaluated from probability curves, matching of LT_{full} for different disturbance amplitudes suggests that the lifetimes of transitional flow structures do not depend on the amplitude of disturbances for $Re \leq Re_{crit}$.

Using the probability curves for three different length pipes shown in figure 5.5, it is possible to obtain the lifetime τ analogous to the 'half-life' notion in radioactive decay. The probability curves is well described by an exponential decay $P(\tau) = a\exp(-\alpha\tau^*)$, where $\tau^* = L/D$ and α is the decay constant, obtained as the slope of the probability curve in a half-logarithmic plot. Hence the lifetime τ is defined as the difference between the τ^* values for $P(\tau^*) = 0.5$ and $P(\tau^*) = 1.0$:

$$\tau = \frac{\ln(2)}{\alpha} \quad (5.2)$$

In figure 5.13(a), the trend line for the probability at $Re = 1900$ is presented. Owing to the dependence of τ on the probability curve and the dependence of the probability curve on the disturbance type and amplitude, it is not expected that τ from different facilities should match. Nevertheless, for the sake of completeness, a comparison of τ^{-1} is provided in figure 5.13(b) with those obtained by Faisst & Eckhardt (2004), Peixinho & Mullin (2006), Hof et al. (2006) and Willis & Kerswell (2007). τ^{-1} in the present investigation

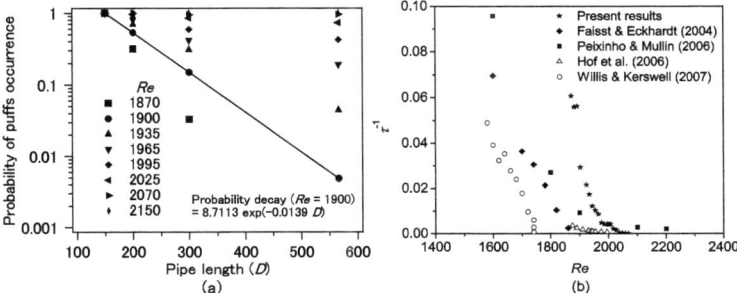

Figure 5.13: (a) Change in probability with L/D and Reynolds number; (b) change in τ^{-1} with increase in Reynolds number compared with previous results

decreases with increase in Reynolds number but it does not tend to become zero. This was caused by the fact that all transitional structures were sustained and thus α tends to become zero but could not possibly be zero, thus τ^{-1} would also never become zero.

5.3 Summary of the results

The lifetime of transitional flow structures in low Reynolds number pipe flow was analyzed by direct measurements of the full-lifetime LT_{full} and the probability of occurrence of puffs by employing an HWA and a pressure transducer. The direct measurement of full-lifetime gave information not only about the lifetime but also about the evolution traces and types in transitional flow structures. Evolution types such as 'just triggered' structures and 'developed' structures appeared depending on the Reynolds number and additionally the existence of bifurcations into 'directly dissipated', 'decayed sustained' and 'sustained' structures at the same Reynolds number was found.

It was shown that LT_{full} was independent of the downstream conditions such as the pipe length and even the amplitude of the disturbances. LT_{full}

of transitional flow structures has various dependencies in different Reynolds number ranges. First, LT_{full} increases exponentially up to $Re \approx 1500$, then it remains constant with increase in Reynolds number between $1500 \leq Re \leq 1550$. At $Re = 1560$, the first indication appeared in the pressure signals that the 'just triggered' structures start to evolve. Clearly bifurcated transitional flow structures, which dissipate in a pipe, were observed for $Re \geq 1880$; one is 'directly dissipated' structures and the other is 'sustained' structures. The sustained structures may decay if the pipe is sufficiently long. 'Directly dissipated' and 'sustained' structures have distinct full-lifetimes; the former has a more or less constant value but the latter has a value that increases with increase in Reynolds number. Due to the sudden occurrence of sustained structures at $Re = 1880$, which was recognized by a sudden large development, this Reynolds number was called the critical Reynolds number. This Reynolds number shows fairly good agreement with theoretically derived results. There was a certain occurrence probability ratio of different transitional flow structures due to their variety in evolutions in pipes observed for $Re \geq Re_{crit}$.

For $1550 < Re < 1880$, LT_{full}^{-1} decreased linearly with increase in Reynolds number, which shows the possibility of transitional flow structures having an infinite lifetime at a finite Reynolds number $Re \approx 2300$. As the probability of split puffs starts to increase at $Re \approx 2300$, as shown in section 4.3, it was suggested that puffs favor breakdown into split puffs before they acquire an infinite lifetime. As a conclusion, the following possible evolution routes for transitional flow structures in low Reynolds number pipe flows were suggested with corresponding possible Reynolds numbers:

1. 'just triggered' structure ($Re < 1500$)

2. start of the evolution of 'just triggered' structures into 'developed' structures ($1500 < Re < 1550$)

3. evolution into 'developed' structures and 'directly dissipating' structures ($1560 \leq Re \leq 1880$)

4. occurrence of 'sustained' structures and increase of their probability ($1880 \leq Re \leq 2080$)

5. breakdown of puffs into split puffs or slug-like puffs ($2325 \leq Re \leq 2680$).

Chapter 6

Conclusions

Laminar to turbulent transitions in pipe flows were extensively investigated with a special test facility that was developed. The facility and also the way in which the experiments and measurements should be carried out to study the present topic are described in detail.

First, the natural transition and transitions triggered by ring obstacles in pipe flows were investigated as preliminary research work. There the applied test rig was verified to function correctly and in well-controlled manner. The characteristics of those transitions were demonstrated with the results obtained by measuring the time variation of the longitudinal velocity on the axis of the pipe outlet. The ring obstacle was set first at the pipe inlet and the dependence of the ring obstacle height on the critical Reynolds number was shown with comparisons, where interesting nonlinear behavior observed at $Re \approx 2000$.

The next set of experiments was carried out in which the triggering location was fixed downstream from the pipe inlet, where the flow had reached fully developed laminar flow. The findings suggested the following:

1. The wall fence obstacle height, normalized with the pipe diameter, to trigger the laminar-to-turbulent transition of a pipe flow changes with Reynolds number as $\sim (Re)_{crit}^{-\frac{1}{2}}$.

2. On plotting the experimental findings as $h^+(Re_\tau)$, it is observed that $h^+ = $ constant. The amplitude perturbation, h^+, is independent of Re_τ. This can be deduced by similarity considerations.

Then an iris diaphragm system, which triggers the flows deterministically to create repeatable transitional flow structures, was employed for the further investigations on the development of puffs and slugs in pipes. The information was obtained by measuring the time variation of the center line velocity at the pipe outlet using different lengths of pipes. In a first set of experiments, the development of slugs was studied and the results showed how the slugs expand along the pipe, as a consequence of differences in the propagation velocities of the front and back edges of the slugs. Second, the development of puffs was investigated with different pipe lengths taking the Reynolds number into account. At fixed Reynolds number, at some distance from the entrance, the structures of puffs and slugs seem to be independent of the nature of the disturbances which created them. The results impressively showed how disturbances turn first into puffs, then splitting puffs and slugs. At lower Reynolds numbers, puffs form and remain nearly unchanged along a pipe, i.e. they showed the characteristics of localized structures. However, when the Reynolds number of the flow is increased, the puffs are characterized by one or more splittings and, for sufficiently high Reynolds numbers, the puffs merge into slugs through puff splitting and start to expand as slugs do in the pipe flows. The results obtained provided a very detailed morphology of the equilibrium puffs and the mechanical aspects of puff splitting and slugs depending on the Reynolds number.

To investigate the dissipation phenomena of transitional flow structures in low Reynolds number pipe flows, the lifetime of transitional flow structures was analyzed by direct measurements of full-lifetime and the probability of occurrence of puffs, employing an HWA and a pressure transducer. The full-lifetime was defined as the time between the iris diaphragm operation and the complete dissipation of transitional flow structures in the pipe. First, the full-lifetime change with increase in Reynolds number was determined, which indicated different trends of full-lifetime in various Reynolds number range. It was shown that full-lifetime at certain Reynolds number is independent of the downstream conditions, such as the pipe length, and even the amplitude of the disturbances. At $Re = 1560$, the first indication appeared in the pressure signals that the 'just triggered' structures start to evolve. Then, a clear bifurcation of transitional flow structures dissipating in a pipe was observed at $Re \geq 1880$: one is 'directly dissipated' structures and the other is 'sustained' structures, having a different full-lifetime. Thus a different way of analysis was applied to describe the full-lifetime of transitional flow structures at $Re \geq 1880$, showing the rate of bifurcated full-lifetime values with the full-lifetime results. For $1550 < Re < 1880$, LT_{full}^{-1} decreases linearly with increase in Reynolds number, which leads the possibility of transitional flow structures having an infinite lifetime at a finite Reynolds number $Re \approx 2300$. As the probability of split puffs starts to increase at $Re \approx 2300$, it was suggested that puffs favor breakdown into split puffs before they acquire an infinite lifetime. The following possible evolution routes for transitional flow structures in low Reynolds number pipe flows are suggested with corresponding possible Reynolds numbers:

1. 'just triggered' structure ($Re < 1500$)

2. start of the evolution of 'just triggered' structures into 'developed' structures ($1500 < Re < 1550$)

3. evolution into 'developed' structures and 'directly dissipating' structures ($1560 \leq Re \leq 1880$)

4. occurrence of 'sustained' structures and increase in their probability ($1880 \leq Re \leq 2080$)

5. breakdown of puffs into split puffs or slug-like puffs ($2325 \leq Re \leq 2680$).

It is suggested that more experimental studies are needed to deepen the insight into the flow behavior in the regime of the laminar to turbulent transition of pipe flows. The theory of traveling waves and the possibility that chaos is organized around them (see, e.g., Kerswell (2005) and Eckhardt et al. (2007)) is an important development in understanding the transition. Even then, more numerical investigations are probably needed for a complete understanding of the laminar to turbulent transition of pipe flows. Also, the transitional flows warrant more investigations in order to show the dependence of the transition on the pipe diameter and compliance with many results that are available in the literature.

Chapter 7

Outlook for future investigations

In previous chapters, the dynamic characteristics of transitional flow structures that develop and evolve in a pipe at various Reynolds numbers were described in detail. Several observations in the introduced investigations suggested that the study of the interior of transitional flow structures was a key issue to describe the laminar to turbulent transition in pipe flows in more detail. For example, in section 4.3, the gross characteristics of the interior of puffs and slugs were shown by measuring u' and Tu values from the velocity signal u. The results implied e.g. that although most features of puffs and slugs are different from each other, their back edge showed in principle the same characteristics in Tu, i.e. both puffs and slugs had a similarity in part of their internal structures.

To gain a deeper insight into the interior of transitional flow structures, a new set of measurements was devised by employing an HWA of x-wire probe to measure the Reynolds stress anisotropy of slugs. The results of the investigation provided some useful suggestions; however, further investigations are required to lead to concrete conclusions. Hence the investigation of anisotropy-invariant measurements is introduced in this chapter as an outlook to show a possible direction for further investigations on transitions further. In section 7.1, a general introduction to anisotropy-invariant mapping is briefly presented. The method and the results of the conducted measurements are presented in section 7.2. A short conclusion for this chapter is given in section 7.3.

7.1 Anisotropy-invariant mapping

Anisotropy invariant analysis is briefly introduced here for the sake of completion.

The continuity and Navier-Stokes equations for a viscous incompressible fluid are the following:

$$\frac{\partial u_k}{\partial x_k} = 0, \quad \frac{\partial u_i}{\partial t} + u_k \frac{\partial u_i}{\partial x_k} = -\frac{1}{\rho}\frac{\partial p}{\partial x_i} + \nu \frac{\partial^2 u_i}{\partial x_k \partial x_k}, i,k = 1,2,3 \quad (7.1)$$

where u_i is the instantaneous velocity, t is time, x is a Cartesian coordinate in a direction i, ρ is the density, p is the pressure and ν is the kinematic viscosity. By introducing the conventional method of separating the instantaneous velocity u_i and p into the mean flow values $\bar{u}_i$ and $\bar{p}$ and fluctuations u'_i and p':

$$u_i = \bar{u}_i + u'_i, \quad p = \bar{p} + p' \quad (7.2)$$

One then obtains the Reynolds decomposed equation:

$$\frac{\partial u_k}{\partial x_k} = 0, \quad \frac{\partial u_i}{\partial t} + u_k \frac{\partial u_i}{\partial x_k} = -\frac{1}{\rho}\frac{\partial p}{\partial x_i} + \nu \frac{\partial^2 u_i}{\partial x_k \partial x_k}, i,k = 1,2,3 \quad (7.3)$$

Then, applying ensemble averaging, one obtains the decomposed equation for the mean flow (Reynolds equation) as follows:

$$\frac{\partial \bar{u}_k}{\partial x_k} = 0, \quad \frac{\partial \bar{u}_i}{\partial t} + \bar{u}_k \frac{\partial \bar{u}_i}{\partial k} = -\frac{1}{\rho}\frac{\partial \bar{p}}{\partial x_i} - \frac{\partial}{\partial x_k}\overline{u'_k u'_i} + \nu \frac{\partial^2 \bar{u}_i}{\partial x_k \partial x_k}. \quad (7.4)$$

One can also obtain the decomposed equation for the fluctuations by subtracting equation 7.4 from equation 7.3:

$$\frac{\partial u'_k}{\partial x_k} = 0, \quad \frac{\partial u'_i}{\partial t} + \bar{u}_k \frac{\partial u'_i}{\partial k} + u'_k \frac{\partial \bar{u}_i}{\partial x_k} = -\frac{1}{\rho}\frac{\partial p'}{\partial x_i} + \nu \frac{\partial^2 u'_i}{\partial x_k \partial x_k}. \quad (7.5)$$

In the derivation of the above equations, it was assumed that the fluctuations are much smaller than the corresponding quantities of the mean flow values:

$$u'_i \ll \bar{u}, \quad p' \ll \bar{p} \quad (7.6)$$

and they satisfy the continuity and Navier-Stokes equations. One multiplies the second equation of equation 7.4 by ρ:

$$\rho\frac{\partial \bar{u}_i}{\partial t} + \rho \bar{u}_k \frac{\partial \bar{u}_i}{\partial k} = -\frac{\partial \bar{p}}{\partial x_i} - \frac{\partial}{\partial x_k}(\rho\nu\frac{\partial^2 \bar{u}_i}{\partial x_k \partial x_k} + \rho \overline{u_k u_i}). \tag{7.7}$$

The third term on the right-hand side of equation 7.7 is called Reynolds stress and written in the form $\tau_{ij} = \rho \overline{u_k u_i}$. To describe the turbulence as a whole, it is necessary to understand the Reynolds stress.

Lumley & Newman (1977) and Lumley (1978) found that the state of turbulence can be characterized by the anisotropy values. The anisotropy of a flow can be derived from the Reynolds stresses $\tau_{ij} = \rho \overline{u_k u_i}$ by subtracting part from τ_{ij} and normalized with $\tau_{ss} = \rho \overline{u_s u_s}$. This leads to the non-dimensional anisotropy tensor

$$a_{ij} = \frac{\overline{u_i u_j}}{2k} - \frac{1}{3}\delta_{ij}, \tag{7.8}$$

where $k = \frac{1}{2}\overline{u_s u_s}$ is turbulent kinetic energy and δ_{ij} is Kronecker's delta. The tensor a_{ij} has three scalar invariants:

$$a_{ij} = 0, \quad II_a = a_{ij} a_{ji}, \quad III_a = a_{ij} a_{jk} a_{ki}. \tag{7.9}$$

By cross-plotting II_a and III_a, the state of turbulence in a flow can be displayed with respect to its anisotropy. If the scalar invariants II_a and III_a are evaluated for the case of a two-component turbulence (one component of the velocity fluctuations is negligibly small compared with the other two), this leads to

$$II_a = \frac{2}{9} + 2III_a. \tag{7.10}$$

Doing the same for the axisymmetric turbulence (two components are equal in magnitude) yields

$$II_a = \frac{2}{3}(\frac{4}{3}III_a)^{2/3}. \tag{7.11}$$

If equations 7.10 and 7.11 are cross-plotted, they define a narrow region called the anisotropy-invariant map as shown in figure 7.1. Lumley (1978) mentioned that all physically realizable turbulence has to lie within this small region.

The boundaries of the invariant map describe the limiting states of turbulence. Isotropic turbulence is found at the lower corner of the map (the

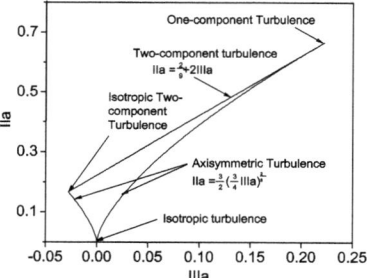

Figure 7.1: Anisotropy triangle presentation

origin in figure 7.1) where $II_a = III_a = 0$, hence the anisotropy is zero. The left branch of the map ($III_a > 0$) describes axisymmetric turbulence in which one component of the velocity fluctuations is smaller than the other two. In contrast, the axisymmetric turbulent on the right side ($III_a < 0$) is characterized by one fluctuating component which dominates the other two. The remaining boundary line on top of the map is the limiting case of a two-component turbulence (equation 7.10). The corner point on the left-hand side of the anisotropy-invariant map represents a turbulence state which consists merely of one fluctuating component.

Hence the bounding states of turbulence of the invariant map are well defined. Based on the results of numerical simulations and experimental investigations on such kinds of turbulent flows following the bounds of the map, Jovanović (2004) was able to formulate complete turbulence closure for all unknown correlations in the transport equations of the turbulent stresses and for all limiting states of turbulence.

7.2 Reynolds stress anisotropy measurement of slugs

The test rig described in chapter 2 was employed to measure the Reynolds stress anisotropy of slugs. The iris diaphragm system was employed at the pipe inlet and triggered the flow periodically. The amplitude of the disturbance, i.e. closing height and lapse time of the iris diaphragm, were kept constant throughout the measurements. An x wire probe was employed for hot wire velocity measurement to obtain two components of turbulence simultaneously. The HWA measurement was carried out at the $L = 8$ m pipe outlet on the center line, hence the flow which was measured was assumed to be axisymmetric. In the present investigation, slugs were chosen for presenting the Reynolds stress anisotropy analysis because they were transitional flow structures which occur at high Reynolds number and thus deterministically, as anisotropy-invariant measurement requires a number of repeated measurements. Over 150 realizations were taken to obtain the Reynolds stress anisotropy invariants. The Reynolds number was chosen as $Re = 12500$ for the present investigation, so that slugs appeared which were highly repeatable.

Figure 7.2 shows a slug realization of an axial velocity at $Re = 12500$, measured at the $L = 8$ m pipe outlet. The Reynolds stress anisotropy invariants are plotted in figure 7.3. The points ABCD in the anisotropy map shown in the figure 7.3 correspond to the points indicated in figure 7.2.

Figure 7.3 depicts the clearly tendency of change in the invariant while laminar to turbulent (A to B) and also turbulent to laminar (C to D) transitions occurred.

The experimental results verified the axisymmetric nature of the transitional flow with large variations of anisotropy starting from its maximum value, which correspond to the one-component state, towards the isotropic state and vice versa during laminar to turbulent transitions.

7.3 Conclusions of the anisotropy measurement of slugs

The conclusion that is drawn from the presented results is that anisotropy-invariant mapping has a good feasibility to describe the interiors of transitional flow structures. The experimental results verified the axisymmetric nature of the transitional flow with large variations of anisotropy starting from

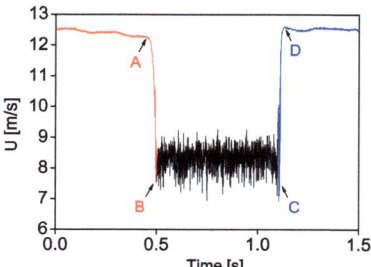

Figure 7.2: Typical slug velocity time profile at $Re = 12500$ showing points A to C for presenting anisotropy-invariant values

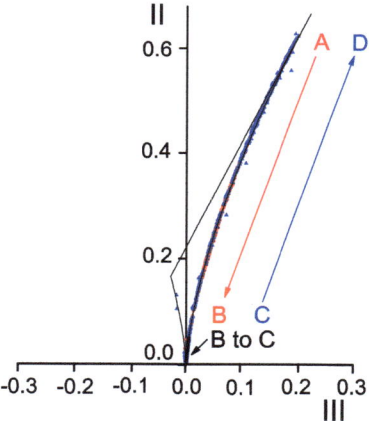

Figure 7.3: Reynolds stress anisotropy-invariant value plots for laminar to turbulent and turbulent to laminar transitions in pipe flows

its maximum value, which corresponds to the one-component state, towards the isotropic state and vice versa during laminar to turbulent transitions. Jovanović (2008) mentioned that this is a verification of the possibility of predicting laminar to turbulent transitions by means of anisotropy-invariant analysis.

References

BANDIOPADHYAY, P. R. 1986 Aspects of equilibrium puff in transitional pipe flow. *J. Fluid Mech.* **163**, 439–458.

BARNES, H. T. & COKER, E. G. 1905 The flow of water through pipes. *Proc. Roy. Soc. London* **74**, 341–356.

BRUNN, H. H. 1995 *Hot-Wire Anemometry.* Oxford University Press.

DARBYSHIRE, A. G. & MULLIN, T. 1995 Transition to turbulence in constant-mass-flux pipe flow. *J. Fluid Mech.* **289**, 83–114.

VAN DOORNE, C. W. H. & WESTERWEEL, J. 2007 Measurement of laminar, transitional and turbulent pipe flow using stereoscopic-PIV. *Exp. Fluids* **42**, 259279.

DOU, H. S. 2006 Mechanism of flow instability and transition to turbulence. *Int. J. Non-Linear Mech.* **41**, 512–517.

DRAAD, A. A., KUIKEN, G. & NIEUWSTADT, F. T. M. 1998 Laminar-turbulent transition in pipe flow for Newtonian and non-Newtonian fluids. *J. Fluid Mech.* **377**, 267–312.

DURST, F. 2008 *personal communication.*

DURST, F., HEIM, U., ÜNSAL, B. & KULLIK, G. 2003 Mass flow rate control system for time–dependent laminar and turbulent flow investigations. *Meas. Sci. Technol.* **14**, 893–902.

DURST, F., RAY, S., ÜNSAL, B. & BAYOUMI, O. A. 2005 The development lengths of laminar pipe and channel flows. *J. Fluids Eng.* **127**, 1154–1160.

DURST, F. & ÜNSAL, B. 2006 Forced laminar to turbulent transition of pipe flows. *J. Fluid Mech.* **560**, 449–464.

ECKHARDT, B. & MERSMANN, A. 1998 Transition to turbulence in a shear flow. *Phys. Rev. E* **60**, 509–517.

ECKHARDT, B., SCHNEIDER, T. M., HOF, B. & WESTERWEEL, J. 2007 Turbulent transition pipe flow. *Annu. Rev. Fluid Mech.* **39**, 447–468.

FAISST, H. & ECKHARDT, B. 2003 Traveling waves in pipe flow. *Phys. Rev. Lett.* **91-224502**, 1–4.

FAISST, H. & ECKHARDT, B. 2004 Sensitive dependence on initial conditions in transition to turbulence in pipe flow. *J. Fluid Mech.* **504**, 343–352.

HOF, B., VAN DOORNE, C. W. H., WESTERWEEL, J. & NIEUWSTADT, F. T. M. 2004 Experimental observation of nonlinear pipe. *Science* **305**, 1594–1597.

HOF, B., VAN DOORNE, C. W. H., WESTERWEEL, J. & NIEUWSTADT, F. T. M. 2005 Turbulence regeneration in pipe flow at moderate Reynolds number. *Phys. Rev. Lett.* **95-214502**, 1–4.

HOF, B., JUEL, A. & MULLIN, T. 2003 Scaling of the turbulence transition threshold in a pipe. *Phys. Rev. Lett.* **91-244502**, 1–4.

HOF, B., WESTERWEEL, J., SCHNEIDER, T. M. & ECKHARDT, B. 2006 Finite lifetime of turbulence in shear flows. *Nature* **443**, 59–62.

JOVANOVIĆ, J. 2004 *Statistical Dynamics of Turbulence*. Springer Verlag, Berlin.

JOVANOVIĆ, J. 2008 *personal communication*.

JOVANOVIĆ, J. & PASHTRAPANSKA, M. 2004 On the criterion for the determination transition onset and breakdown to turbulence in wall-bounded flows. *J. Fluids Eng.* **126**, 626–633.

KERSWELL, R. R. 2005 Recent progress in understanding the transition to turbulent in a pipe. *Nonlinearity* **18**, 17–44.

KERSWELL, R. R. & TUTTY, O. R. 2007 Recurrence of travelling waves in transitional pipe flow. *J. Fluid Mech.* **584**, 69–102.

LINDGREN, E. R. 1969 Propagation velocity of turbulent slugs and streaks in transition pipe flow. *Phys. Fluids* **12**, 418–425.

LUMLEY, J. L. 1978 Computational modeling of turbulent flows. *Appl. Mech.* **18**, 123–176.

LUMLEY, J. L. & NEWMAN, G. 1977 The return to isotropy of homogeneous turbulence. *J. Fluid Mech.* **82**, 161–178.

MELLIBVOSKY, F. & MESEGUER, A. 2006 The role of streamwise perturbations in pipe flow transition. *Phys. Fluids* **18**, 074104.

MESEGUER, A. & TREFETHEN, L. N. 2003 Linearized pipe flow to Reynolds number 10^7. *J. Comput. Phys.* **186**, 178–197.

MESETH, J. 1974 Esperimentelle Untersuchung der Übergangszonen zwischen laminaren und turbulenten Strömungsgebieten in intermittenter Rohrströmung. *Mitteilungen aus dem Max-Planck-Institut für Strömungsforschung und der Aerodynamischen Versuchsanstalt* **58**, 1–114.

MULLIN, T. 2008 *personal communication.*

MULLIN, T. & PEIXINHO, J. 2006 Transition to turbulence in pipe flow. *J. Low Temp. Phys.* **145**, 75–89.

PEIXINHO, J. & MULLIN, T. 2006 Decay of turbulence in pipe flow. *Phys. Rev. Lett.* **96-094501**, 1–4.

PEIXINHO, J. & MULLIN, T. 2007 Finite-amplitude thresholds for transition in pipe flow. *J. Fluid Mech.* **582**, 169–178.

PFENNINGER, W. 1961 In boundary layer suction experiments with laminar flow at high Reynolds numbers in the inlet length of a tube by various suction methods, in *Boundary Layer and Flow Control*, ed. by G. V. Lachman, Pergamon Press, Oxford, pp. 961–980.

REYNOLDS, O. 1883 An experimental investigation of the circumstances which determine whether the motion of water shall be direct of sinuous, and the law of resistance in parallel channels. *Philos. Trans. R. Soc. London, Ser. A* **174**, 935–982.

ROTTA, J. 1956 Experimenteller Beitrag zur Entstehung turbulenter Strömung im Rohr. *Ing-Arch.* **24**, 258–281.

RUBIN, Y., WYGNANSKI, I. J. & HARITONIDIS, J. H. 1980 Further observations on transition in pipe, *Proc. IUTAM Symp. Stuttgart, FRG*, Springer, Berlin, pp. 19–26.

SCHILLER, L. 1934 Neu berichte zur Turbulenzentwicklung. *Zeitschrift für Angewandte Mathematik und Mech. (ZAMM)* **14**, 36–42.

SCHNEIDER, T. M., ECKHARDT, B. & VOLLMER, J. 2007a Statistical analysis of coherent structures in transitional pipe flow. *Phys. Rev. E* **75**, 066313.

SCHNEIDER, T. M., ECKHARDT, B. & YORKE, J. A. 2007b Turbulence transition and the edge of chaos in pipe flow. *Phys. Rev. Lett.* **99-034502**, 1–4.

SHAN, H., ZHANG, Z. & NIEUWSTADT, F. T. M. 1998 Direct numerical simulation of transition in pipe flow under the influence of wall disturbances. *Int. J. Heat Fluid Flow* **19**, 320–325.

SREENIVASAN, K. R. 1982 Laminarescent, relaminarizing and retransitional flows. *Acta Mech.* **44**, 1–48.

TREFETHEN, L., CHAPMAN, S., HENNINGSON, D., MESEGUER, A., MULLIN, T. & NIEUWSTADT, F. 2000 Threshold amplitudes for transition to turbulence in a pipe. *http://arXiv.org/abs/physics/ 0007092* .

ÜNSAL, B. 2006 *personal communication.*

WEDIN, H. & KERSWELL, R. R. 2004 Exact coherent structures in pipe flow: traveling wave solutions. *J. Fluid Mech.* **504**, 333–371.

WILLIS, A. P. & KERSWELL, R. R. 2007 Critical behavior in the relaminarization of localized turbulence in pipe flow. *Phys. Rev. Lett.* **98-014501**, 1–4.

WILLIS, A. P. & KERSWELL, R. R. 2008 Coherent structures in localized and global turbulence in pipe flow. *Phys. Rev. Lett.* **100-124501**, 1–4.

WILLIS, A. P., PEIXINHO, J., KERSWELL, R. R. & MULLIN, T. 2008 Experimental and theoretical progress in pipe flow transition. *Philos. Trans. R. Soc. London, Ser. A* **366**, 2671–2684.

WYGNANSKI, I. J. & CHAMPAGNE, F. H. 1973 On transition in a pipe. Part 1. The origin of puffs and slugs and the flow in a turbulent slug. *J. Fluid Mech.* **59**, 281–351.

WYGNANSKI, I. J., SOKOLOV, M. & FRIEDMAN, D. 1975 On transition in a pipe. Part 2. The equilibrium puff. *J. Fluid Mech.* **69**, 283–304.

ZANOUN, E. M. 2007 *personal communication.*

Kurzfassung

In der vorliegenden Dissertationsarbeit wird ein klassisches Problem der Strömungsmechanik nämlich die laminar-turbulente Transition in der Rohrströmung untersucht und die gefundenen neuen Erkenntnisse werden zusammengefasst. Untersuchungen zur Transition im Rohr haben eine lange Ge-schichte, die zurück geht bis Reynolds (1883) und seither laufend fortgeschrieben wurde. Dennoch gibt es immer noch eine ganze Reihe offener Fragen, die auf eine Klärung warten. In der Einleitung werden diese Fragen auf der Basis einer kurzen Literaturübersicht herausgearbeitet. Für die anschließend durchgeführten Experimente wurde ein Versuchsaufbau verwendet, der von Durst & Ünsal (2006) erstellt und mit einigen Modifikationen an die geänderte Aufgabenstellung angepasst wurde. Der Versuchsstand, der aus einem Massenstromkontroller, einem Strömungskonditionierer, einem Messrohr, einer Einrichtung zur Auslösung der Strömungstransition und den Messgeräten bestand, wird im Detail beschrieben. In einer einleitenden Studie wurde die natürliche Transition sowie die Transition, die durch ringförmige Hindernisse ausgelöst wird, betrachtet. Es zeigte sich eine deutliche Abhängigkeit der Transitionsauslösung von der Höhe des ringförmigen Hindernisses sowie von der Reynoldszahl, wobei die Reynoldszahl definiert ist als $Re = u_{bulk}D/\nu$, mit u_{bulk} der Durchflussgeschwindigkeit, D dem Rohrdurchmesser und ν der kinematischen Viskosität des Fluids. Wie bekannt ist, können zwei unterschiedliche Arten von Strömungsstrukturen bei der laminar-turbulenten Transition in Rohren beobachtet werden, nämlich Puffs und Slugs, wie sie erstmals von Wygnanski & Champagne (1973) eingeführt wurden. In der vorliegenden Arbeit wurde die Entwicklung von kontrolliert erzeugten Puffs und Slugs während ihres Transports entlang der Rohrlänge bei unterschiedlichen Reynoldszahlen untersucht. Dabei konnte auch die

Umwandlung von Puffs zu Slugs über ein Puff-splitting oder aber das Abklingen der Puffs bei niedrigen Reynoldszahlen beobachtet werden. Der Zeitraum zwischen der Erzeugung der Puffs und ihrem Verschwinden wurde als 'full-lifetime' definiert und konnte mit einem Drucksensor direkt gemessen werden. Auf der Basis dieser Lebenszeitmessungen konnten die möglichen Entwicklungen der Strömungsstrukturen, die bei der laminar-turbulenten Transition von Rohr-strömungen niedriger Reynoldszahl auftreten, diskutiert und weitgehend geklärt werden. Für das weitergehende Verständnis der Übergangsphänomene in Rohrströmungen wurde die Anisotropie der Reynoldsspannungen in der Slugs-struktur mit Hilfe der Hitzdraht-Anemometrie vermessen. Die Ergebnisse zeigen, dass das Reynoldsspannungsanisotropiemodell in der Lage ist, den Übergang gut vorherzusagen.

Acknowledgements

It is my great pleasure to acknowledge all people who supported me while carrying out my Ph.D. thesis work at the Institute of Fluid Mechanics of Friedrich Alexander University Erlangen-Nürnberg (LSTM-Erlangen).

First of all, I thank very much Prof. Dr. Dr. h.c. Franz Durst, who is my Doktorvater. He has known me since 2001 through an introduction by Prof. Hajime Fujimoto, who was my bachelor thesis supervisor at Doshisha University in Kyoto. The purpose of my move to Erlangen, Germany, was not only to study as a master course student, but also to conduct research works under his supervision. I am very happy that I have such a Doktorvater, who has continuously encouraged and supported me and given me proper guidances and supervision during the whole of my Ph.D. thesis work.

I am very grateful to Dr. Özgür Ertunç and Dr. Bülent Ünsal who obtained their Ph.D. degree at LSTM-Erlangen, for their help, suggestions and encouragement. A large number of scientific discussions held with them were really fruitful, and finally reflected well in the completion of my Ph.D. thesis work. I owe special thanks to PD. Dr. Jovan Jovanović, head of the turbulence group at LSTM-Erlangen, for his scientific guidance.

I appreciate the generous support of Prof. Dr. Antonio Delgado, who is the head of LSTM-Erlangen, and I am grateful to all of my colleagues and the technicians at LSTM-Erlangen for their helpful advice, meaningful discussions and the good times which we spent together. I thank my parents, brother and friends who always supported me during my stay in Germany.

Erklaerung

Ich versichere, dass ich die Arbeit ohne fremde Hilfe und ohne Benutzung anderer als der angegebenen Quellen angefertigt habe und dass die Arbeit in gleichen oder aenlicher Form noch keiner anderen Pruefungsbehoerde vorgelesen hat und dieser als Teil einer Pruefungsleitung angenommen wurde. Alle aus fuehrungen, die woertlich oder sinngemaess uebernommen wurden, sind als solche gekennzeichnet.

Erlangen, den 30.01.2009

VDM Verlagsservicegesellschaft mbH

Die VDM Verlagsservicegesellschaft sucht für wissenschaftliche Verlage abgeschlossene und herausragende

Dissertationen, Habilitationen, Diplomarbeiten, Master Theses, Magisterarbeiten usw.

für die kostenlose Publikation als Fachbuch.

Sie verfügen über eine Arbeit, die hohen inhaltlichen und formalen Ansprüchen genügt, und haben Interesse an einer honorarvergüteten Publikation?

Dann senden Sie bitte erste Informationen über sich und Ihre Arbeit per Email an *info@vdm-vsg.de*.

Sie erhalten kurzfristig unser Feedback!

VDM Verlagsservicegesellschaft mbH
Dudweiler Landstr. 99
D - 66123 Saarbrücken

Telefon +49 681 3720 174
Fax   +49 681 3720 1749

www.vdm-vsg.de

Die VDM Verlagsservicegesellschaft mbH vertritt

Printed by Books on Demand GmbH, Norderstedt / Germany